Ron Stites

Technology Development

Also of interest

Process Engineering.
Adressing the Gap between Study and Chemical Industry
Michael Kleiber, 2020
ISBN 978-3-11-065764-7, e-ISBN (PDF) 978-3-11-065768-5,
e-ISBN (EPUB) 978-3-11-065807-1

Process Development.
An Introduction for Chemists
Jerry Carr-Brion, 2022
ISBN 978-3-11-071786-0, e-ISBN (PDF) 978-3-11-071787-7,
e-ISBN (EPUB) 978-3-11-071791-4

Technoscientific Research.
Methodological and Ethical Aspects
Roman Z. Morawski, 2019
ISBN 978-3-11-058390-8, e-ISBN (PDF) 978-3-11-058406-6,
e-ISBN (EPUB) 978-3-11-058412-7

Sustainable Process Engineering.
Gyorgy Szekely, 2021
ISBN 978-3-11-071712-9, e-ISBN (PDF) 978-3-11-071713-6,
e-ISBN (EPUB) 978-3-11-071730-3

Process Machinery.
Commissioning and Startup - An Essential Asset Management Activity
Fred K. Geitner, Ronald G. Eierman, 2021
ISBN 978-3-11-070097-8, e-ISBN (PDF) 978-3-11-067415-6,
e-ISBN (EPUB) 978-3-11-067427-9

Ron Stites

Technology Development

Lessons from Industrial Chemistry and Process Science

DE GRUYTER

Author
Ron Stites
Stites and Associates, LLC
175 South Main Street
Brighton CO 80601
USA
ron@tek-dev.net

ISBN 978-3-11-045171-9
e-ISBN (PDF) 978-3-11-045163-4
e-ISBN (EPUB) 978-3-11-045196-2

Library of Congress Control Number: 2022931115

Bibliographic information published by the Deutsche Nationalbibliothek
The Deutsche Nationalbibliothek lists this publication in the Deutsche Nationalbibliografie; detailed bibliographic data are available on the Internet at http://dnb.dnb.de.

Cover image: Ron Stites
Typesetting: Integra Software Services Pvt. Ltd.
Printing and binding: CPI books GmbH, Leck

www.degruyter.com

Contents

Chapter 1
Introduction

This book is a much-expanded version of my 2015 eBook, *The Art, Science, and Discipline of Technology Development* [1]. The changes are extensive enough to differentiate this from a "second edition." Nevertheless, the motivation for writing on this topic is still very much the same.

When I first started the eBook, I thought I might call it something like "Managing Research and Development." I decided against that for two reasons:

- Few managers believe they are managing research and development ("R&D")
- Few companies recognize any R&D activities going on in their organization.

In other words, few would be interested in such a book because it has no bearing on what they think they do. Quite to the contrary, most managers deal with elements of R&D quite frequently. Furthermore, nearly all modern companies are interested in new technologies and often develop them. Nevertheless, R&D has many negative connotations. Business managers view R&D as a "study without end." They get this impression from how academics approach R&D. That is not the kind of R&D I am talking about here. Hence, throughout this book, I use the term "Technology Development" not as a contrivance but as a shift in thinking that emphasizes controlled processes aimed at definite purposes.

What I teach in this book rarely appears in science or engineering courses – even at graduate levels. Science and engineering courses do not usually include financial valuation topics. It is also rare to see these concepts in business courses. Business courses rarely include serious attempts to teach statistical inference. Hence, nowhere do students learn about the nexus between the financial valuation of technology and statistical inference. It is unclear to me why this is.

I suspect that financial valuation is of little interest in the sciences. That is probably a legacy idea that the "pure sciences" should not be tainted by filthy lucre. To improve the value of the technology we develop, we must overcome that prejudice. I wonder if engineering and business schools struggle more with teaching statistical inference than not valuing it. Students certainly arrive at the university with little background in statistics, and the mathematics of statistics is undoubtedly complex. The problem is probably more about how we "teach" at the university level in America. There is little emphasis on the learning process – especially in higher education.

Most professors are subject matter experts rather than teaching or learning experts. As a result, they are not focused on how well they teach their students or their success after graduation. Instead, professors are rewarded for publishing, getting grants, and dollars brought into the university through patents, stipends, programs, and gifts. The bottom line:

https://doi.org/10.1515/9783110451634-001

> Even with Graduate Degrees, students rarely learn how to think critically, evaluate data statistically, and make good, common-sense decisions. As a result, they are, for the most part, poorly equipped to lead or even engage in Technology Development.

Our society must, however, succeed at Technology Development. We must provide the training and develop the leadership skills needed. If we fail, our country will continue to fall behind in this area. We must look at learning and knowing differently.

In the past 40 years, we have seen a shift from production-based business to knowledge-based business. Peter Drucker anticipated this in his 1959 book, *The Landmarks of Tomorrow* [2]. He predicted the demise of the "Blue Collar Worker" and the emergence of the "Knowledge Worker." Few dispute the reality of this massive shift, especially in the Western world. Critics pine away for the "good old days" when there was little value differentiation between educated and uneducated labor. Those days will never return – not even in underdeveloped countries. Today most of us deal in the arena of knowledge, particularly new and unproven "knowledge."

Unfortunately, our ability to think about knowledge has not kept pace with the need to possess it. We now believe that "knowledge is power," but we are unsure what this means. We are especially baffled by the distinctions between such notions as claims and facts, suspicions and reality, and data and information. Our technical ability to promulgate words and ideas has far outstripped our ability to sort out truth from fiction.

Fifty years ago, this was a bit disturbing. We had a sense that we often were not being told the truth. Nevertheless, we expected truth and, where practical, demanded it. A little-known senator from Missouri, Harry S. Truman, became famous when he exposed and punished fraudulent defense contractors in what became known as the Truman Committee. He eventually became president. In the 1960s, we were up in arms when we learned that President Lyndon Johnson had lied about the Gulf of Tonkin, getting us into the Vietnam War. Hundreds of thousands eventually ended up in the streets protesting the war. In the 1970s, Richard Nixon resigned from office for lying about Watergate. In the 1980s, President Ronald Reagan felt compelled to apologize for misleading the public on the Iran Contra Affair.

Then came the post-deregulation, Go-Go days of the 1990s. Speculation ran rampant, followed quickly by a series of financial failures and crises that started with the Savings and Loan Crisis and continued for almost 20 years with the Dot Com Bubble, the Asian Financial Crisis, the Housing Market Collapse, and the Global Financial Crisis of 2007–2008. After that, we became indifferent to the gulf between fact and fiction. A seminal moment came on January 26, 1998. Since the day Clinton got away with saying "I did not have sexual relations with that woman" on national television, the lies have gotten bigger, and the consequences for lying have diminished.

We saw large corporations recalling millions of products that did not work or were unsafe. We wasted billions of tax and investment dollars on technology schemes that never had a chance of working and government programs that yielded precisely the opposite of the intended result. Business leaders became multimillionaires by selling products and programs that users had to fix or immediately "upgrade." We went to war because of a WMD (Weapons of Mass Destruction) crisis that did not exist. And nothing much seemed to happen to stop the madness. We now seem to be alone, adrift in a sea of lies, deceit, and disinformation. How do we begin to get to the truth about anything?

This book attempts to guide determining "fact" from "fiction." It is not easy, but I believe it to be essential. The book focuses on Technology Development, but the thinking processes discussed here have broader implications. I founded it on the following premises:

- Truth exists to be discovered – it is not manufactured.
- Nothing can be known with certainty.
- No practical knowledge exists outside of experimental evidence.
- Decisions must be made with incomplete knowledge.
- The future is neither completely determinate nor indeterminate – it is influenced by previous actions without being completely determined by them.

If you are looking for certainty, you should pick up something in the religious section of the bookstore. It is not that they are any more correct than anyone else. They just think they are. There is no certainty in this world. When it comes to truth, we are in the uncomfortable position of only improving our confidence through the effort and expense of experimentation. In other words, we must make preliminary guesses at truth and then test them in structured ways.

As we put these preliminary guesses to the test, we can enhance their reasonableness. To this extent, we are "proving" them. We are "proofing" them as one would "proof test" a high-pressure vessel or a gun barrel. Without experimentation to "proof" a claim, it remains unsubstantiated and may have no basis. It might be a myth, a rumor, wishful thinking, or even propaganda, but it is certainly not absolute truth.

The bogus claims, the broken promises, and the wasted resources of the past few decades have caused much of our current economic suffering. This suffering includes our drift toward greater inequality in the distribution of wealth. There is hardly anything more corrosive to economic health as financial risk. Corrupted economic and political systems create a drag on economic progress that is large and insidious. Those "in the know" cheat those "in ignorance" when it comes to buying products and services, investing in companies and retirements, and supporting political candidates. The result is the emergence of an elite oligarchy victimizing and impoverishing the general population – especially the middle class. The tactics employed are propaganda, voodoo science, and appeals to feelings and fears rather than facts. As a result, populist movements arise, and political instability grows.

Sadly, we cannot personally verify everything. Attempting this would leave us hopelessly ignorant. So instead, we must rely on the work and reports of others, whether it is a new way to brew coffee or a change to medical insurance. We must recognize "good work" from "poor work." We must insist that decision-makers ask the right questions and subject processes to review by competent persons without undue conflicts of interest or political pressure.

When we oversee evaluating potential changes – whether for our organization or ourselves – we engage in some form of technology evaluation and development. We must investigate more directly the claims of others and assess their diligence. We must be aware of the many fallacies and traps that will come our way. Some of these will derive from the naive enthusiasm of fools, but an alarming number of them will be deliberate cons perpetrated by a host of charlatans. Deceptions are everywhere, and none so convincing as self-deception.

The purpose of this book is to:
- raise awareness of the need for validation of claims,
- give some of the "science" on how to evaluate claims,
- give some guidance on the "art" of managing evaluation and development processes,
- give some guidance for building organizations that create and sustain a culture where critical evaluation and development of new ideas is an innate discipline, and
- encourage broad participation in improving our thinking and the institutions that touch Technology Development.

This book is for a broad audience engaged in formal Technology Development. These are executives, supervisors, support staff, engineers, and scientists. It is also for everyday citizens trying to make sense of the dizzying array of "scientific" polls and studies that seem to "prove" that the earth is flat and that aliens abducted Elvis. It gives general thought patterns along with some statistical mathematics.

Topics in the book are:
- What is technology? – This is not a book about computer hardware or software. Instead, the book includes examples from a wide variety of technologies and a variety of fields "touched" by Technology Development.
- What is Technology Development? – How do we learn about a specific innovation and measure its value? Why should we care about developing new technologies? What is the role of risk in Technology Development?
- What are some of the basic "assumptions" about Technology Development?
- What are some of the "tools" of Technology Development? – This includes some technical tools like hypothesis testing, but it goes much deeper into evaluation, management, and financing processes.
- Is there a disciplined process for Technology Development? – Where do we start? When do we "give up?" When are we "done?"
- Who should be doing Technology Development? Is this the responsibility of an elite intelligentsia? Are there other roles to be played?
- How should we fund Technology Development? Does one size fit all?

- What should Technology Development look like in the future? How should we train future technologists? Are there better ways to fund Technology Development?
- Conclusions – This is a call to both personal and corporate actions.

Throughout the book, we will be referring to past Technology Development projects. These are projects in which the author was a significant participant. They will not be identified by name because these were private companies. Instead, they will be referenced in the text as examples to illustrate a point.

I hope that the reader will gain a sense of urgency about supporting truth in our society. We need a new "army" of critical thinkers strategically placed to stop the siphoning of precious dollars into the many hair-brained and deceptive schemes that drain the wealth of our nation and erode what little remains of our ethics and morals. It is time to turn on the bright lights of actual, verifiable knowledge. I hope this book will serve as a manifesto in that regard.

Chapter 2
What Is Technology and Technology Development?

Some common misconceptions

Let's first look at a few myths:
- Technology = Computers
- Technology Development = Research = Science
- Technology Development = Engineering = Development
- Technology Development is unpredictable

The highly touted success of Silicon Valley has given the public the impression that Technology Development only deals with computers and computer applications. That is dangerously false. It is naïve to believe that writing software code and managing data is the core wealth-producing activity in the modern world. Silicon Valley is not the only source of creativity and wealth today. Even the "revolution" in electronics is far more complex than just writing code and managing databases. Other engineering and application skills not associated with programming or data mining are becoming more valuable.

Computers and computer programming have made possible an incredibly diverse set of potential applications. Nevertheless, the high-value "breakthroughs" are not in computing chips or data storage. Instead, the significant developments in computers today are being made by mechanical, electrical, aeronautical, and material science engineers, along with economists, political scientists, educators, marketers, quality control professionals, and a dizzying array of experts in other vocational fields. These scientists and engineers are developing the sensors, motors, materials, communication devices, data models, vehicles, logistical control models, roads, teaching methods, and even social and economic theories to apply computers more effectively to real-world needs.

The issues today are more about how computers can gather "appropriate," "real-time" data, and use it to do something in the physical world than it is about making calculations or storing data. High-value activities are now designing and building the sensors to collect the real-time data, establishing which data are appropriate to collect, developing algorithms that improve decision-making, and developing devices to allow the computer to act in the real world.

Another myth is that Technology Development is scientific research done by the "experts" in a university or a National Laboratory. This myth is bunk. It results in hundreds of millions of dollars of waste every year. Some of the worst people to engage in Technology Development are self-proclaimed "scientists." Most "scientists" are devoted to promoting their peculiar "science." If they are a nuclear physicist, they are looking to apply a nuclear reactor to "solve" your early childhood

https://doi.org/10.1515/9783110451634-002

reading problem. Their "solution" is not improving reading among children but figuring out how in the heck a nuclear reactor could do something useful related to early childhood reading. It is difficult to get scientists focused on the technologies needed to solve problems and resist the temptation to develop "their" technology, which often proves to be of little use.

Another myth is that Technology Development is always best done by an engineer or the engineering department. Engineers are frequently "development" averse. They often want to "get things done" rather than "figure things out." Many are neither equipped nor trained to study a problem and hunt for anything that is not an obvious, "off-the-shelf" solution with guaranteed performance specifications. Some have been "programmed" by their career experiences to avoid complex problems and anything that might fail.

As we will see, scientists and engineers can be helpful with Technology Development but are generally not the ones to lead the effort. The action leader must focus on outcomes in an environment where failures are likely. Few Technology Development efforts proceed smoothly from idea to commercialization without serious side trips and false starts. Scientists often lose focus, and engineers lose heart. It takes a different kind of person to keep going as long as it makes sense and "throw in the towel" when the project becomes a "money pit."

And one final myth is that Technology Development is entirely unpredictable. This notion comes from anecdotal tales about how "all" the great discoveries have been pure accidents. This lie is the biggest whopper of them all. We learn nothing accidentally. Even the "great accidents" were careful observations and structured experiments where existing knowledge had been developing for many years. The "discoverer" was alert enough to know that the "accident" was not a purely random, uncontrolled event but an unexpected result that challenged assumptions. Furthermore, they knew enough about what was going on to speculate that this surprising result might have a practical application.

So persistent is this idea that a popular "theory" has grown up around it. That is, the "Infinite Monkey Theorem." The "theorem" goes something like this:

> A monkey banging away randomly at a typewriter for an infinite amount of time would almost surely create all the works of Shakespeare.

The conclusion is that "discoveries" are bound to happen given enough time and effort. Hence, the key is to keep "banging away." This theory is total nonsense. Several different estimates of the time needed for the monkey (or even a universe full of monkeys) to create even "Hamlet" would be so long that human civilization would "almost certainly" be long gone. What's worse, if, by accident, the one monkey would defy all the odds and write a single sonnet, he probably wouldn't realize what he had done and eat the one miraculous copy.

The same applies to "banging away" at Technology Development. Random acts take far too long, and not knowing what you want will likely leave you throwing away the "discovery" that you didn't recognize. Hence, effective Technology Development is a disciplined process. Anyone claiming otherwise doesn't know what they are talking about or is trying to excuse sloppiness and waste.

Types/Classes of Technology

Not all technologies are alike. They differ dramatically in their physical nature, users, and market impact. We can prioritize our activities by keeping in mind the peculiarities of the technology. Although not an exhaustive list of types/classes of technology, we certainly see differences around:

Capitalization:
- Hardware – capital-intensive development and deployment
- Software – labor-intensive development and deployment

Markets served:
- Business to business (B2B)
- Consumer markets (B2C)
- Government markets (defense/security/health care/infrastructure)
- Social/common good (nonprofit deployment)

Impact on existing markets
- Incremental improvements
- Disruptive change

Many new technologies are hardware-oriented. They could be a unique catalyst for cracking crude oil, a new aerospace alloy, a new CPU chip, or a new type of car. However, these technologies are often expensive to develop because of expensive test equipment, facilities, personnel, or raw materials needed during testing. They may also be costly to commercialize. The "minimum viable product" could turn out to be a minimum production capacity out of a billion-dollar plant.

Hardware products can also be risky to develop. Frequently, established competitors are "entrenched in the market." Existing hardware users often have sunk investments and are reluctant to change to a new hardware solution. The newcomer has a high burden of proof that the new hardware solution is worth the risks and costs to the user. Often there are artificial barriers to entry, including government regulations or "industry standards." These could be arbitrary and even capricious barriers that maintain mini-monopolies or obsolete historical suppliers.

Contrast this with a "software" technology. We have a lot of labor needed here, but prototyping, testing, and deploying a software solution rarely require building large production plants. The cost of deploying a new software technology tends to be low – perhaps even too low. Silicon Valley has conditioned us to accept unreliable, inadequately tested software technologies. Nevertheless, this could be changing –especially where security risks are significant.

Developing hardware technology is nothing like developing software technology. The costs, the investors, the users, the deployment channels, and the impact on the market will be different. Hence, it is little wonder that cash-flush software developers struggled when they tried to expand into other markets.

Technologies also differ in the types of markets they will serve. Developers must recognize the wants and needs of the target market to be successful. An essential part of the Technology Development process is creating and documenting features that benefit a target market. It is surprising how frequently technology developers fail to recognize what their target markets want or need. As a result, many Technology Development programs develop "cool" features that customers don't value.

Example 1: Without a Customer Need There Is No Value
A biomass-to-alternative fuels start-up initiated an extensive Technology Development project using a mix of government and private funding. The inventors had proven features of the technology at a small scale. The project's primary purpose was to show potential investors an integrated solution using multiple technologies that converted waste wood into a drop-in alcohol fuel additive. A key challenge was identifying the ultimate customer for the product. Unfortunately, the developers never reached clarity on the target market. Hence, they focused on product features that didn't matter while failing to deal with some key features that turned out to be deal-breakers. Eventually, the developers abandoned the project.

One technique for understanding the market to be served is to "segment" into distinct categories. Each exercise will be unique and should reach as fine a segmentation as is needed to elucidate essential customer needs while identifying large enough segments to support the deployment of the technology into the market. These technology features identified should drive the priorities of any Technology Development process. Below are some broad segments that serve as a starting point:

B2B Markets

Specifications, costs, and risks drive a Business-to-Business (B2B) relationship. Customers are not likely to adopt new technology without clearly demonstrated benefits. These benefits can include reduced costs, improved quality, improved reliability, a higher value to downstream customers, and similar economic benefits. Occasionally, regulatory compliance and public perception can be important contributing factors. In any case, a B2B "sell" requires reliable data to convince rational decision-makers that the benefits of adopting the new technology outweigh the costs and risks.

The B2B approach is one of the most challenging Technology Development paths. It is somewhat rare that start-ups succeed on this path – especially those with hardware-based technology. It usually takes many years of development and can cost millions of dollars. Few start-ups can fund such a long-term, expensive process. However, those who succeed can emerge as the new, multibillion-dollar company. These projects often start as spin-offs from established companies or quickly morph into a subsidiary of a much larger company in an existing industry.

Consumer Markets

On the surface, this appears to be an easier path than B2B, but that could be an illusion. Consumers may adopt new technology very readily, but catching that market is an "art." Steve Jobs of Apple Computer was a master at this, but he was very rare among technologists. Most technologists do not understand consumers and will frequently miss the market very badly. Furthermore, most technologists do not understand nor appreciate the need for sales, marketing, or distribution of consumer products. Start-ups without access to these resources have little chance of success. If they have a great idea, they may lose it to a "copycat" before they can enjoy the benefits.

In any case, developers should collect evidence that consumers will be interested in the new technology. One place to look for guidance is the excellent book, *Contagious, Why Things Catch On*, by Jonah Berger [3]. Dr. Berger lists six features of ideas that "catch on" with consumers. He concludes that most "hot ideas" have two or more of these features. Technology developers would be wise to look at this book and convince themselves that their new technology has a chance to "catch on." Where practical, the Technology Development process should enhance those features to improve consumer interest and acceptance.

Example 2: If Customers Cannot Be Reached There Is No Value

A start-up company developed and tested a portable wine aging device. Nearly 100 consumers tried the device at taste-testing studies held in multiple venues. It became clear that the device worked, but the consumer needed direct support to use the device effectively. The company modified the design so that consumers could join a support group and easily share tips and tricks. Unfortunately, the company could not find an appropriate marketing partner willing to provide the necessary customer report. Rather than launch a product that would likely fail, the company shelved the technology.

Government Markets

Government entities (federal, state, county, city, etc.) are frequently interested in supporting Technology Development. They fund Technology Development through a variety of programs giving grants and loans. Often "matching" funds are required, money the developer must provide. Unfortunately, government entities rarely provide 100% of the funding necessary to fund for-profit companies.

Government entities will have specific reasons for making funds available for Technology Development. These can range from general economic development/recovery programs to programs designed to develop crucial defense technology. They may or may not allow the technology developer to own or control the technology developed. Hence, the technology developer must understand the purpose of the government program and the particulars of any funding agreement. A surprising number of programs will allow the developer to own and control the technology without specific constraints. Other programs have stringent limitations on what the developer can (or must) do with the developed technology. In virtually all cases, the federal government reserves the right to control the technology if the government deems it vital to national security.

The grants and loans that tend to be the most "flexible" promote economic development or recovery. These programs assist in creating sustainable businesses that will support local economies. These programs rarely put restrictions on the use or ownership of any intellectual property (IP) developed during the program. Most programs of this type require that the technology developer show how the technology's development will lead to a sustainable business or industry. Therefore, the Technology Development program must develop and document features that support that path to economic sustainability.

The most restrictive programs are those that support national defense (including cyber security). Falling somewhere in between the two extremes are a wide variety of technical and social priorities. For example, complex technical issues related to healthcare, energy, transportation, communication, agriculture, mining, or similar fields with broad impact might receive unrestricted funding because solving these issues would positively impact the economy. In contrast, other, less crucial technologies might be set aside for educational institutions, nonprofits, or minority-owned businesses.

In all cases, it is the responsibility of the technology developer to ensure adequate alignment between the policy goals of the government entity and the business model or business needs of the developer. Furthermore, the technology developer must ensure that key "success" factors for the government entity and the developer will be satisfied. Alignment between government and developer goals will significantly impact the development process and priorities. This alignment will impact the Technology Development program in several ways:

- Perceived misalignment of goals will reduce the probability of the developer receiving funding.
- Actual misalignment of goals not resolved during negotiations will result in conflicts with potentially severe financial and legal consequences.
- Remedies for the actual misalignment of goals will frequently result in a project that is "successful" from the perspective of the government entity but does little to create value for the technology developer.

Example 3: "Learning" Is Not Always Valuable
A company developing a new catalyst for turning syngas (a mix of hydrogen and carbon monoxide gases) into a transportation fuel additive won a Cooperative Research and Development Agreement (CRADA) with a National Laboratory. The idea was to research the catalytic mechanism to solve specific issues such as premature deactivation and poor selectivity. Unfortunately, the developer and the National Lab wrote the CRADA too loosely. As a result, the developer and the Lab learned much about the reaction mechanism but little about the catalyst. Hence, the "successful" contract did little to advance the development of the technology.

Hence, government funding may or may not be helpful for the technology developer. It depends on:
- how closely government funding goals align with the business plans for the developing company and
- how freely the developing company can apply the technology developed to obtain a return on the development cost (i.e., "monetize" the technology developed).

In some cases, the monetization of the technology can be follow-on sales to the funding government agency. This case frequently happens with weapons and security developments. Nevertheless, monetization will often be through using or licensing the technology for sale in the private sector. This approach adds another complication to the valuation of a government-developed technology. There is no guarantee that the technology will be a commercial success.

Example 4: The Green Fleece
A start-up company used a mix of government and private funding to develop a more efficient solar panel. The technology used an innovative design and incorporated more efficient (and expensive) materials. The technology was successful, and the company used other government and private funds to build a highly automated production plant to manufacture the new panels. Unfortunately, existing, well-established solar panel makers responded to the new entrant by lowering prices below the break-even point of the new entrant. As a result, the company went bankrupt, and the US government lost nearly a half-billion tax dollars.

Social/Common Good

Historically, the arena of "doing good" was the exclusive purview of "charitable" organizations. However, things began to change with the Tax Reform Act of 1969. This Act created the 501(c)3 nonprofit organization, redefining nonprofit organizations. As a result, nonprofit organizations have grown dramatically in number, size, and scope over the past few decades. Many state and local governments have become keenly aware of this and have included the growth and health of nonprofits as factors in their economic development strategies. Hence, governments began to take an interest in promoting nonprofits for financial reasons.

At the same time, the lines between for-profit and nonprofit began to blur. Some nonprofits were allowed to "lobby," while others were not. Universities became significant players in patenting technologies developed using government grants. For-profit organizations began to demonstrate "social awareness." Some nonprofit organizations set up related, for-profit organizations to sell spin-off products and services. Government entities started awarding grants and loans to for-profit companies if they would "partner" with nonprofit institutions to meet social justice criteria. There became a dizzying array of regulations and practices that made "interesting" bedfellows.

Against this chaotic backdrop exist at least three major Technology Development sectors that seem to persist in the social benefit arena:
- "MedTech,"
- "EdTech," and
- "GreenTech."

Of course, there are other sectors we could add, but these are among the "hottest" going right now. These are the "hottest" ones now because they have the public's attention, driving government spending. Healthcare, education, and environmental sectors have long been the nonexclusive purview of government and nonprofit funding. This focus will likely remain so in the foreseeable future.

Unfortunately, government funding is fickle. What was hot for the last administration isn't necessarily what the next administration thinks is even worthwhile. Furthermore, the approach to solutions for social benefit issues is rarely consistent. When priorities change, funding follows. Hence, "chasing" the latest funding fad is a high-stakes game that couples high risk with low profit. Add to this the difficulty in assessing the monetary value of the "social benefits," and we have the makings of an asylum where a $1,000 toilet seat "makes sense."

The rational analysis that we propose in this book may not directly apply to the "social benefit/common good" sectors of Technology Development as things now stand. Today, government entities award contracts and the press dispenses accolades for many reasons, rarely based on direct, rational valuations. Nevertheless, these subjective "goods" have tangible, significant objective costs. Therefore, policymakers should use the techniques described in this book to aggregate the development costs and the probability of success whenever trying to justify "investing" in new technology. Even when the main driver is a "social benefit," the "investors," whomever they are (and that might be the taxpayer), deserve to know the objective costs and the risk of failure. When policymakers fail to objectively "count the costs," they create an environment conducive to fraud and moral hazard.

Financial resources available for investment in Technology Development are limited. Wasting financial resources in high-risk schemes can have significant wealth generation consequences. Investing investment dollars in low-return, social-benefit activities that have low or no return will reduce the average earning power of capital, reducing economic growth. Low economic growth can exacerbate economic inequalities and create additional social needs. When done in the extreme, excessive funding of social benefit activities can cause hyper-inflation that endangers the long-term health of an economy.

When dealing with Technology Development designed to meet social benefit criteria, it is critical to show that the technology has a high probability of producing the desired effect without causing unintended consequences. Unfortunately, it be can be challenging to investigate all possibilities, especially complications arising from balancing multiple, subjective evaluation criteria. As a result, many well-meaning attempts have failed to deliver the expected benefits.

Example 5: The Corn Ethanol Debacle

The corn ethanol industry is an example of a social benefit effort gone awry. The initial driver for government subsidies was to provide a "renewable," nontoxic, low-carbon dioxide footprint source of oxygen-containing additives for motor gasoline. These oxygen-containing additives would help reduce carbon monoxide emissions when burned in a passenger vehicle. A subsidiary goal was to provide a low-cost octane enhancer to replace tetra-ethyl lead. This program created a new industry – the corn ethanol plant. Unfortunately, this was never a very economically sustainable solution. Even today, after many years of development and persisting (though reduced) subsidies, plants can barely make ends meet. The low-carbon dioxide footprint hope has turned out to be a mirage. At the same time, increased demand for corn artificially inflated corn prices, and agricultural water use has increased disproportionally. The upshot of all of this is that the corn ethanol project is not the "poster child" of the environmental movement and continues primarily due to the support from the farm lobby. It is somewhat ironic that two significant and indispensable sources of profit for big corn ethanol producers are income from farm futures and the sale of carbon dioxide produced during fermentation to the beverage industry.

Impact on Existing Markets

Not only can technologies be characterized by the markets they impact, but we may also characterize them by their impact on those markets. There are two basic types of impact types:

- incremental improvements and
- disruptive change

In his landmark book, *The Innovator's Dilemma* [4], the late Clayton M. Christensen introduced the idea that some technologies are incremental improvements ("Sustaining" was the term Christensen used). In contrast, others are disruptive technologies that lead to disruptive changes in the market. The book's main point was that established companies frequently miss these disruptive changes and fail. Existing companies often focus exclusively on incremental improvement in their current products and services. As a result, they are blind to those fledging, underdeveloped technologies that will change the market and make them obsolete. The excellence of their ability to manage incremental improvements often became an impediment to recognizing newer threats.

Companies frequently miss emerging technologies because:

1. the emerging technologies develop in different fields outside the incumbent company's natural "radar,"
2. the initial introduction of a new, partially developed technology often shows poorer initial features than the well-developed, existing technology, and
3. implementing a new technology has immediate costs in transitioning operations and sales operations to accommodate the latest technology.

Developing and commercializing these two different types of technologies can be quite different. Frequently, incremental improvements are made "in-house" by existing companies or their most trusted vendors. As a result, they developed under great secrecy and became "trade secrets" used as a competitive edge in the existing market.

Example 6: Keeping the Technology to Yourself

A medium-sized steel plant developed a new lubrication system for machines that made steel grinding balls. The mining industry used hardened steel balls to crush rocks and minerals on a large scale. This product was one of the most profitable steel products made by the plant. The lubrication system depended on replacing hardened tool steel and oiled brass sliding surfaces with nylon bearing surfaces lubricated with pressurized oil. The innovation was a simple retrofit that resulted in a surprising improvement in the wear life of the sliding surfaces. It was developed internally by the plant maintenance department. The company considered patenting the new system but decided to keep it a trade secret and implement the innovation only within its other US plants.

Example 6 illustrates how an incremental improvement might have significant economic benefits for the user but have little open market IP value. In this case, the market was too narrow and too far afield of the company's core business to make patenting and licensing worthwhile. Instead, the company used the innovation internally to make more profit in its grinding media business. Frequently, the inventor(s) of new, internally developed systems receive no additional remuneration for the invention. Many innovations of this type never come to the attention of the public.

Example 6 also illustrates some of the difficulties for third parties developing incremental improvements. Companies often hire consulting and engineering firms to work on specific problems and create innovative solutions. However, rarely is the consulting or engineering company allowed to own the technology developed. Hence, they are usually just "hired hands" that get paid for their time and expertise. There are, of course, exceptions to this rule. Occasionally, engineering firms with long-term relationships with clients will work out IP sharing arrangements that make sharing both inventorship and economic benefit more equitable.

Example 7: Recognizing the Serendipitous Event

A start-up company developing a catalyst to make mixed alcohols stumbled across a simple method to modify an existing catalyst into one that does a fair job of making mixed alcohols. The technique was not particularly successful for making mixed alcohols but had potential in other processes. The start-up company applied for and received a patent. The company included the patent in its total company portfolio. When the start-up decided to abandon its mixed alcohol efforts, it sold part of its portfolio of patents to a catalyst manufacturing company. The catalyst company resold the patent for a significant, undisclosed gain to a chemical company that saw considerable potential in the method.

Example 7 illustrates what can happen when a company discovers something the value of which the company does not recognize or cannot monetize. In this case, the company realized that there was value in the discovery (that is why they obtained a patent), but the start-up did not have access to the catalyst market. Without access to a customer base, the company could realize no value. Hence, the company sold the patent at little value as part of a whole portfolio. The buyer, however, had access to the catalyst market and knew where to turn to recognize value for the patented method.

The start-up company attempted to make incremental improvements on catalysts for a particular purpose. Unfortunately, the company could not benefit from that discovery, but others who knew the catalyst market made the bulk of the profit from the research. The examples show that technology type often determines the beneficiary of the study. Unless the developer has a realistic plan for monetizing the research, investments are in vain. Technology type will also affect the development process itself.

In the case of incremental technology, assessing value is easier to judge since a market already exists. Details about potential cost reductions or feature improvements will be easier to evaluate and turn into financial data. However, the criteria for acceptance of the innovation could be very demanding. Potential customers will often have considerable upfront acceptance costs. Furthermore, when this type of development is being done by a third party, assessing competing interests inside potential users can play havoc on development plans. The third-party developer rarely knows enough about the potential customer's business to evaluate the customer's sunk costs adequately, opportunity costs, and regulatory constraints. What seems a significant improvement to the developer could be a nonstarter for the potential customer.

In the case of disruptive technology, assessing financial benefits to potential customers can be very difficult. First-generation technologies are often highly flawed. The inability to estimate what the second- or third-generation improvements might look like complicates valuation. It is little wonder that existing businesses frequently miss the market "turn." They are so focused on making today's products and services "work" that they have little inclination to even think about "new methods," especially those that initially have "warts." Those developing disruptive technology must "sell the vision" to those who are rarely visionary.

One final warning is needed before we move on. Those managing Technology Development must keep in mind that all inventors believe that their invention is disruptive. The inventor believes their vision will revolutionize any industry it touches. Furthermore, it will "sell itself." Nothing could be farther from the truth. Developers must design programs that create the data needed to "sell" the idea to potential customers. Those potential customers will likely be reluctant to adopt the new technology. Hence, managing Technology Development is more than just doing experiments and collecting objective features and benefits. The technology must overcome multiple objections and be adopted. Therefore, we will say again and again that Technology Development is a planned process that requires both science and art.

The Benefits of Technology Development

The "reputation" of Technology and Technology Development waxes and wanes. After breakthrough developments like the invention of the Haber ammonia process, the Salk polio vaccine, or the smartphone, we tend to get heady about our view of technology. However, a war, famine, or natural disaster always seems to follow shortly, which shakes our confidence, and we go back to our nervous wondering. There can be little doubt that Technology Development and population growth are correlated. But is there causality? Does Technology Development make us healthier and happier so that our population grows, or do the challenges of population growth spur Technology Development so that we can survive?

Whether we are Malthusian, Boserupian, or something in between, it seems to me that Technology Development is an indispensable part of modern life. Much of the world population would quickly perish without many of our current agricultural and energy technologies. We would plunge into abject poverty and chaos. The world's economies would promptly shrink as production and transportation would grind down to a snail's pace. It would be difficult to argue that technology has not contributed to worldwide wealth in significant ways. Hence, it seems clear that technology contributes to wealth, but what about wealth distribution?

In his international bestselling book, *Capital in the Twenty-First Century* [5], Dr. Thomas Piketty investigates wealth's growth and distribution. Piketty notes that national growth in wealth has two components. One is demographic (i.e., population growth), and the other is purely economic (i.e., wealth growth per capita). It is the latter that determines the standard of living. Economic growth was low and surprisingly stable until the industrial revolution. Those technological innovations increased economic growth dramatically. Hence, technology has been primarily responsible for the dramatic increase in wealth.

Nevertheless, Piketty also notes that the distribution of that new wealth has been very uneven since the 1700s, with the last few decades being especially concerning – so much so that the dislocations predicted by David Ricardo and Karl

Marx might become a reality. He concludes that the impact of technology on wealth distribution has been erratic. Although technology has contributed to the value of "human capital" (e.g., income flowing to knowledge workers), it has also enhanced the value of the financial capital needed to develop and implement technology.

Piketty's analysis of long-term wealth distribution focuses on the split between the benefits received from economic growth to the capital providers compared to those obtained by providers of expertise/labor. He posits a "first law of capitalism" as follows:

$$\propto = r X \beta \tag{1}$$

where α is the share of the national income received by capital providers, r is the rate of return on capital, and β is the ratio of the total national capital to annual national income.

When r is high, capital providers take a large share of the annual national income to use that capital. This disproportionate share contributes to capital accumulation (i.e., growth of wealth) for the holders of money and property. In simple terms, the rich get richer at a faster rate compared to workers. Hence, a high rate of return on capital tends to create additional inequality in wealth distribution.

Piketty introduces one additional concept and a historical observation that have significant meaning. He notes that wealth for those not owning capital comes primarily from national income and that wealth growth for this group comes mainly from the change in the national income, which he calls "g." He then reviews the history of r and g and notes that periods of increased wealth inequality correlate with when r exceeds g. In other words, inherited wealth tends to grow faster than wealth accumulated through work. He terms this as a "fundamental force for divergence" in the distribution of wealth. He further notes that those with wealth can often manipulate the price of capital upward (particularly with real estate, petroleum, commodities, etc.), making capital dearer and pressuring "r" ever upward. He sees these factors as conspiring to make capitalist societies trend toward less economic equality over time.

Piketty addresses the notion that technological improvements could lead to greater equality because knowledge workers are necessary for those improvements. The argument runs that these knowledge workers could demand high wages and grow their wealth by participation in national income. He proposes this as a potential source of increased wealth distribution – especially if a path to knowledge work was made widely available. In other words, if higher education were widely accessible, a way to knowledge work would be open to many.

Nevertheless, in a more recent book, *Capital and Ideology* [6], Piketty provides a scathing critique of education in the USA (see pages 513–540). He praises US dominance in primary and secondary education up to around WWII. Nevertheless, he is especially critical of current US higher education, where knowledge workers try to enhance their skills at a high personal cost. He concludes that the US system of higher education caters to the rich (domestic and foreign) while leaving the poor behind and the middle class to struggle:

> We come now to one of the principal challenges that social-democratic societies must face today, namely, the issue of access to skills and training, especially Higher Education. Property is important, but education has also played a central role in the history of inequality regimes and the evolution of social and economic inequalities both within and between countries. Two points deserve particular attention. First, throughout much of the twentieth century, the United States has held a significant lead in education over Western Europe and the rest of the world. This US advantage dates back to the early nineteenth century and beyond, and it explains much of the large gas in productivity and standard of living that one observes through most of the twentieth century. In the late twentieth century, the United States lost this lead and witnessed the appearance of a new stratification with respect to education: significant gaps in educational investment separated the lower and middle classes from those with access to the most richly endowed universities. [6], page 513

I believe that Piketty is correct in his assessment of US higher education but is incomplete. The issues are more profound than just funding and financial structuring. The problems also include the quality of higher education in the USA. Professor Jacques Berlinerblau does a savage critique of higher education in his 2017 book, *Campus Confidential* [7]. This highly entertaining and somewhat vulgar book is a must-read for parents and students looking at higher education. It exposes the inability of the hierarchical, medieval model of higher education to deliver quality instruction. Berlinerblau's diatribe leaves the reader wondering how knowledge workers would ever get the knowledge, technical competence, and critical thinking skills they need to succeed. The bottom line is that we cannot expect US higher education to deliver equitable wealth distribution. Instead, our higher education reinforces the primacy of the wealthy in an increasingly expensive, inefficient, pompous, and self-serving academic system.

The details of how to fund and organize higher education are outside the scope of this book. Nevertheless, this book has a lot to say about effectively managing Technology and Technology Development. The technical entrepreneurs (and financiers, for that matter) need to learn a variety of skills to understand and critique the pronouncements of the scientists and engineers. Furthermore, scientists and engineers need to know how technology is correctly valued. It does not do for these two different groups to misunderstand and distrust each other. It makes no sense to fund technology projects that are not both technically and financially feasible. Funding dumb projects that sound good causes the cost of capital to skyrocket and forces required return on investment upward. Current "rules of thumb" for venture capital run something like "triple my money in five years with no technical risk." When developing technology, the required internal rate of return jumps between 35% and 50%! These are unsustainable levels.

Assessing the risks and rewards of technical projects is a problematic, multidisciplined endeavor. As we will see, it requires technical/scientific/engineering evaluations along with market assessment, financial management, project management, and a host of other disciplines. Doing this well is the best way to reduce the required return on capital for Technology Development. With less risk and fewer abject failures, a smaller return on capital becomes feasible. Hence, the providers of expertise (i.e., the knowledge workers) can take a higher percentage of the increased income.

If we make practical Technology Evaluation more broadly available, we may drive down the cost of capital while increasing the distribution of economic benefits. Hence, teaching a broad range of individuals the skills needed to participate in the evaluation process is an effective way to make the financial benefits from new technology available to a wider variety of persons. That is one of the purposes of this book.

Example 8: Teaching Priorities and Nanyang Technological University

One of the more progressive technology training programs I ever saw was the Renaissance Program run by Nanyang Technological University (NTU), Singapore. In 2015, I had the distinct pleasure of working with six Chemical Engineering Interns on a research project in the USA. What surprised me was that these students did not know as much chemistry as I expected. What they did know, however, was how to do research and how to do financial valuations. The NTU program taught business, financial, management, and critical thinking skills before introducing the technical nuts and bolts of science and engineering. These are skills that US universities never seem to get around to teaching.

Chapter 3
Some Basic Axioms of Technology Development

Technology Development is a process. It is not magic. It is not a dependence on miracles. It is incremental but not "steady." There are fitful "breakthroughs" and bouts of bad luck. Hence, some "luck" is involved, but it is a poker game and not a chess match. Before we dive into the Technology Development process, I should give you some insight into my prejudices. I call them "prejudices" not because there is no evidence to support them, but rather, I don't want to take on the burden to "prove" them to the skeptical. Instead, I have found them helpful "axioms" that I now take for granted.

Axiom 1 – Nothing Is Completely Predictable

The first of these is that nothing is entirely predictable. However, we can obtain reasonable probabilities on almost anything with careful effort. The key is to make good, rational decisions that maximize our likelihood of success, guard ourselves against the probability of failure, and let the chips fall where they may. To be good at Technology Development, one must admit the reality of uncertainty and manage it. For more on living in a world of uncertainty in a general sense, I recommend the excellent book, *Thinking in Bets, Making Better Decisions When You Don't Have All the Facts* [8], by Annie Duke. She was a very successful professional gambler who turned to management consulting. If there were ever examples of making decisions without all the "facts," professional gambling and Technology Development would have to be among them.

Technology Development includes two significant tasks. The first is assessing the probabilities for costs and benefits and judging the desirability of the status quo. The next is taking steps to change those probabilities and evaluating the desirability (and perhaps necessity) of investing in further actions to change the status quo. When thought of in this light, it becomes clear that Technology Development is:
- an iterative process,
- a mix of technical and business problems, and
- a combination of subjective and objective issues.

Because nothing is entirely predictable, the future is not predetermined – you must hang around for the results. Furthermore, it is impractical to control the future although you can influence it. Until something happens, it is never certain, and no amount of effort or money can make a specific outcome inevitable. Although we can take steps to reduce uncertainty, the cost and effort to do so go up without bounds. As we improve our chances for success, we learn that each increment of

https://doi.org/10.1515/9783110451634-003

improvement costs more than the one before. Hence, the "art" of decision-making includes balancing the incremental cost of improving our certainty with the risk of being wrong. We will always have to decide to act without perfectly knowing the outcome. Winners choose decision points that make for "good bets" over the long term and avoid "betting the farm" on "sucker" deals.

It is worthwhile to dwell on predeterminism for a moment – especially for our culture. Americans are deeply conflicted on this philosophical point. Our culture is steeped in Calvinism and influenced by its predeterminism. Hence, many take for granted that the future is fixed even if they don't know why. On the other hand, many Americans tend to be optimistic and react strongly against predeterminism. They take for granted that the future is completely open and that anything is possible. Both positions are demonstrably false. Leaders need to be aware of firmly held prejudices about the future when selecting team members for projects.

Axiom 2 – Collaboration Can Be a Four-Lettered Word

The second axiom is that "teams" are more valuable than "collectives." Leaders should design groups around well-developed tasks and goals. These will be much more effective in getting things done than self-directed groups that can wander off track. For this reason, I use the term "collaborate" very sparingly and focus much attention on who is "leading" the team and what the decision-making processes are. Later, we will talk about executives, technologists, and technology entrepreneurs. I will give a brief description here with more details later.

> Note: I will be using the term "Executive" in the sense of being a traditional "business type" that does not know the technology and may not be well trained in Technology Development. This type of Executive will need the help of others (like a trusted Assistant described below) to accomplish their mission. There is, however, an emerging management type that could be called a Technical Entrepreneur. These individuals have technical and business training/experience – often an MBA. This management type is generally more successful with Technology Development.

Executives, technologists, and technology entrepreneurs are classes of individuals with specific views of who they are and their roles in the organization. The executives are the persons with formal decision-making authority and frequently determine the organization's financial success. Technologists are the scientists and engineers providing the "brainpower" for Technology Development. Technologists often feel disconnected from the organization and are more loyal to their field than the business. The tension between these two groups is described in chapter 18 of *The Personnel Management Process* [9] by Wendell French.

The technology entrepreneur is a more recent synthesis between these two types. It is the technology entrepreneur that this book hopes to create. We can sometimes see an individual technology entrepreneur, but that is rare. Instead, we often see a team of individuals cooperating to reach common goals.

Example 9: When Technology Development Is in Silos

Growing up, I wanted to be a researcher. I read biographies of Thomas Edison, Alexander Graham Bell, Nikola Tesla, Guglielmo Marconi, and others. When I was about to graduate from KU with my BS in chemistry (1973) and go on to Iowa state to get a Ph.D. in quantum chemistry, I decided to look at what it would be like to be a researcher in the industry. My uncle, L. S. (Bob) Carsey (a partner at the prestigious Houston law firm of Fulbright & Jaworski), set me up to visit several research chemists in the Houston area. I was stunned that so many of these bright researchers were miserable. With few exceptions, they were frustrated by the "business types" running the programs. The "business types" often assigned them to useless or hopeless projects. So I did not go to Iowa state. Instead, I worked for a while and any job I could get. I eventually landed an environmental chemistry job and stayed there for several years. Later in my career, after getting an MBA in finance and accounting, I was able to get into industrial research, first in management and later in Technical Development.

Axiom 3 – Money Is the Final Unit of Measure

The third axiom is that, in a business setting, the value of almost everything can and should be reduced to a money value, even if only approximate. This principle is fundamental when group economic interests are at stake. Admittedly, reducing decisions, relationships, obligations, and intangible property to monetary value can be challenging, especially in a large organization. It could require some sophisticated managerial accounting to assess costs and benefits when sharing resources fairly. Nevertheless, the effort usually pays big dividends.

Here, I am explicitly exempting from consideration the personal scruples and prejudices of the individuals in the group. Individuals can and should make personal value judgments on subjective bases such as "right," "wrong," "ethical," "noble," "desirable." That is part of being human. Individual, subjective choices should be respected and accommodated where practical. Nevertheless, leaders should not impose those individual choices as unchallenged scruples for the group. Imposing scruples is the beginning of tyranny and squeezes out of the organization the kind of diversity that is needed to avoid "group-think."

A corollary to this axiom is that, in the long run, the cost of Technology Development must ultimately be economically sustainable. The value of the technologies developed must exceed the cost of producing those technologies, including capital and a premium for the risks taken. Private companies that ignore this rule go broke. Government programs that ignore this rule are not only embarrassed in the public eye, but they also misappropriate scarce investment dollars, driving up the cost of capital and eventually slowing economic growth. Driving up the cost of capital has the double negative effect of reducing general wealth and contributing to the concentration of wealth.

This axiom also points the way toward the methods of valuing technology. Reasonable valuations should be hardnosed internal rate of return valuations supported by reasonable estimates of potential profitability. Detailed business plans provide the support required for those reasonable estimates. Many asinine "investment models" for valuing Technology Development exist based on arbitrary valuations of patents, previous investment, early market penetration, and so on. These aren't worth the paper they are written on. Reasonable valuations come from what customers will pay for a product or service. Hence, any decent plan needs to identify potential markets and work backward, determining development, commercialization, production costs, revenue, and the timing of costs and revenue. This process needs to be detailed enough to support the investments requested and identify the risks involved. If a customer's need is not satisfied by the technology, it will have no value. No one should invest in developing that technology.

We are talking here about Technology Development that creates usable technology that pays out for stakeholders. There are many examples where venture capitalists (VCs) "funded" dumb projects and made money doing it. A typical strategy for a VC is to put in some "seed money," hype the deal to attract other investors (including government entities), get most of their money out, and leave everyone else "holding the bag." The VC has made money on their investment by about the third round of funding. If the deal turns out to be dumb and craters at that time or later, the VC has made money from the other investors/stakeholders. The result is a transfer of wealth from investors and taxpayers to the VC and not an increase in wealth.

Axiom 4 – Everything Is Urgent

The fourth axiom recognizes that time may begin as a friend but always become an enemy. Nothing is constant over time. No plan. No assessment. No market condition. No investor commitment. Nothing. The passage of time incessantly erodes value.

Furthermore, the impact of time is neither linear nor easy to predict. Sudden and surprising changes can and do happen. Every plan is becoming obsolete. Customer needs are changing, and competitors are changing the marketplace. Changing economic conditions can turn the most stalwart strategic investor into a "vulture capitalist" desperate for a quick buck. Hence, Technology Development is not something to be done at a leisurely pace. A critical competitive factor is developing better and faster than the competition. The "need for speed" frequently leads to an optimal organizational size for the Technology Development team. Smaller teams can often make decisions faster. Small groups, however, may suffer from inadequate resources to do the hard development work. Therefore, leadership must strike a compelling, competitive balance.

Axiom 5 – Size Matters

The rules for the very small are not the same as those for the very large. Not in size and not in number. Not in physics and not in finance.

A small family business does not have the same risk/reward/funding rules that a Fortune 500 has. The small family business has almost no access to institutional investors and only minimal access to bank funding. Fortune 500 companies have limited access to many government programs designed for small businesses. Hence, all business plans need to consider the size of the business and the prevailing opportunities and constraints.

Additionally, discrete (countable) events may not follow the same mathematical models as measurable quantities. Our standard techniques for statistical modeling (normal distribution, expected values, standard deviations, etc.) may not apply when we care about countable events like individual failures, customer choices, correct decisions made by managers, and the like. Furthermore, simple calculations are often inadequate when our options are limited and consequences significant. We may be delighted to risk a dollar on the throw of a die when we can win ten bucks on calling the correct number, but who would play Russian Roulette all day to win $100 on each spin of the cylinder?

Axiom 6 – The Quest for Knowledge Requires Intuition and Deduction

The final axiom is that the path to knowledge is not strictly objective. We don't know what we don't know. Hence, we must frequently resort to intuition, guesses, and even prejudices as places to start our development and serve as our bases for evaluating progress. We frequently have no better places to start than a poor first guess. If we assess those intuitions objectively through careful and honest testing, we can often make quick progress toward understanding – even when our starting assumptions were very wrong. To do this, we must always keep in mind that our current knowledge contains many assumptions, some of which may turn out to be flawed. We must remain humble about what we think we know and open to other possibilities that can come along. Those possibilities may impeach our most fundamental assumptions and make our current "model" completely obsolete.

Example 10: A New Corn Ethanol Technology

A start-up company applied an existing technology to the corn ethanol fermentation process. The technology injected high-pressure steam into the corn mash to assist in the breakdown of the corn starch to free sugar. Testing verified that the steam injection created free sugars much faster than enzymes alone. The research group concluded that corn ethanol plants could reduce the cost of expensive enzymes used to turn cornstarch into free sugar. The start-up invested heavily into a commercialization program that included setting up trials at existing corn ethanol plants. Unfortunately, additional careful research revealed that the yeast that fermented the free sugars into alcohol was "poisoned" by too much free sugar in the mash. The yeast died prematurely, and ultimate ethanol yields suffered. The slower, enzymatic method provided free sugar to the yeast at an optimal rate, leading to significantly more ethanol production. The "new" information made the steam injection system obsolete. The company abandoned the project.

Chapter 4
The Science of Technology Development

So, where do we start? The best place to begin is at the end – the customer. We should begin with a customer's need and develop a technology to satisfy that need. Surprisingly, this isn't how it usually happens. Often, Technology Development starts with a "cool idea." This idea may come from our tendency to fund many lame-brain ideas in academia, where researchers are probably most separated from real-world market needs. In any case, when presented with a new technology, the first question should be, "Who cares?" The second question should be, "How much (as in how much money) do they care?" If no one cares, we are done with our Technology Development project. We should write a simple report and go have a beer.

Markets First

For a technology to have value, a customer must value it. That value needs to be translatable into money – at least a reasonable estimate. It will cost money to develop this technology; hence, if customers will not pay for development, implementation, and production, this is a losing proposition. The Technology Development landscape is littered with the carcasses of solutions looking for problems.

This initial evaluation is never a simple thing to do but is one of the most important. It immediately starts the discussion about who is the potential customer. However, that discussion should go deeper into:

- How does the customer perceive this need?
- Precisely what is our product or service?
- What alternatives do the customers have?
- How will we convince customers to adopt/buy our solution?
- What will it cost us?
- How much will the customer pay?

The inventor (especially from academia) is reluctant to subject their "cool idea" to such pecuniary assessment, so don't expect to be applauded for such an odious analysis. Nevertheless, this is the type of discussion that is needed to even do the first cut on, "Is this 'cool idea' something of value?" It is important to remember that Technology Development resources are scarce and that "cool ideas" are a dime a dozen. In fact, since the passing of the Bayh-Dole Act in 1980 that allowed universities, nonprofits, and small businesses to patent federally funded research, junk research and garbage patents in the USA have exploded. On the other hand, great ideas are scarce. It takes time to evaluate technology and determine its merit. Therefore, we should reject obvious losers as quickly as practicable and move on to better prospects.

https://doi.org/10.1515/9783110451634-004

The detail required to answer these questions depends on the "investment" required. Answering tough questions is always an iterative process. The initial request might be an informal question, "Would you look at this?" Inevitably, the research becomes more expensive and more detailed as time goes on. A cursory review of the market potential could justify a few thousand dollar feasibility study. However, only a thorough, detailed market analysis could justify building a pilot plant costing millions.

Developers often overlook how the technology should be "monetized." Some questions might be:

1. Should we develop the technology for others to use or implement?
2. Should we try to manufacture the product or supply the service? Directly or through others?
3. Should we partner with others in the field?

Answers to these questions will also drive the Technology Development strategy. It requires an objective analysis based on many factors, including:

- internal constraints like size and funding of the Technology Development group,
- time available to develop the technology,
- technology type,
- market conditions, and
- cost and time of development.

What should emerge from a market analysis is the beginning of a living document that becomes a research and development (R&D) plan. It should contain many elements of a business plan specific to the technology under development – especially how it satisfies a customer's need. It should also include the details (as they emerge) of the nuts and bolts of the investor/stakeholder plans and the experimental designs discussed later. Finally, developers should continuously update this R&D plan as the Technology Development proceeds.

Investors/Stakeholders Second

Although this may come as a surprise to some, the next step in the process is to begin thinking about funding – including external funding. Of course, this requires building a budget around the Technology Development project. Developers must consider how much money they need when they need it and who will provide the funds. It also requires building a financial strategy around making money for potential investors. Especially critical is thinking through the required return to investors. The return necessary will depend on who the investor is, their motives for investing, and how they view the risks of investing.

There are at least eight sources of funding for Technology Development. These include (in order of early to late-stage):

- Personal savings and "sweat equity" (for small and very early projects)
- Family and friends (again usually small and early projects)
- Government grants and loans (for technologies of interest to specific agencies)
- Bank loans (for going concerns with assets and income)
- Strategic partners (other companies with an interest in the technology)
- Angel investors (private companies or individuals with interest in the technology)
- Venture capitalists (sharks with interest in making obscene amounts of money quickly)
- Institutional investors (larger firms that will invest large sums into profitable businesses for expansion)

We will talk more about these funding sources later, but now is the time to map out the path to commercial success. Each group has different interests and needs. They have varying requirements on the rate of return, participation in the process, and information needed to make an investment decision. The financing plan should also be a living document that is updated as new information comes available about the markets, the costs, the timing, and the potential internal rates of return (IRR).

It is also essential to enlist critical stakeholders and supporters and "neutralize" critical detractors and enemies. Positioning the technology in the marketplace takes time and effort. It is never too early to identify these tasks and plan to take appropriate actions throughout the project. The stakeholders, supporters, and enemies can include regulators, suppliers, wholesalers, politicians, trade groups, unions, professors, news pundits, bloggers, competitors, and the public. Some technologies draw favorable attention and are an easy "sell." Other technologies are hard to explain or understand and need trusted supporters to get them adopted. Still others have negative impressions that developers must actively address in long-term open discussions and negotiations. Developers should document these activities in the Technology Development plan and include the resources needed to implement these activities. Hence, the Technology Development plan frequently includes studies, trials, and experiments designed to address safety, environmental, and even social justice impact concerns.

The executive needs to keep these living documents updated in their head in a condensed, "30-second" "elevator speech." The executive should be ready to explain the status of the Technology Development project, including the best estimates of the costs, benefits, "ask" (i.e., how much money is needed), and IRR. Many deals are made or lost in a few minutes of an "at-the-water-cooler" or "at-the-cocktail-party" meeting with key influencers, decision-makers, and potential investors.

Go/No-Go Decision Process Third

Creating a "go/no-go" decision mechanism early in the Technology Development process is vital. As mentioned above, every step in a Technology Development project is more expensive than the one before. The first iteration of a market/investor analysis can be something of a mind experiment. If that passes the "smell" test and the Technology Development project still seems plausible, a "Go" to the next phase will probably result in the first serious investment of time, money, and ego. This decision should be memorialized, explicitly stating the reasons for a "Go" decision. More importantly, developers should record any "contingencies" or "hypotheticals" around that decision. The Technology Development process will settle these suppositions and unknowns.

Example 11: The Catalyst That Wasn't

An alternative fuel company was developing a homogeneous catalyst to turn syngas into diesel fuel. It is always a challenge to separate the homogeneous catalyst from the products. This problem was well known to the company to be a "deal-killer" if not solved. Over time, this issue got "lost" among the many research priorities of the company. Some 20 years of investor-supported work passed as the company seemed to be working toward commercialization. When a new research team member noticed that this vital issue was unresolved, all hell broke loose. Within a matter of weeks, the company went into a death spiral.

Furthermore, the developers must schedule the next "go/no-go" decision point. It should be clear when the "Go" decision will be reviewed and by whom. Unresolved "deal killers" should be considered at each "go/no-go" review. Projects tend to gain a life of their own. They can become both impossible to complete and impossible to abandon. Therefore, it is imperative developers have as much clarity as possible about when a project is failing. For the good of the organization and players, developers must frequently review and objectively evaluate progress toward commercialization.

For industrial Technology Development, the executive will be the best qualified to perform the "go/no-go" evaluation. They will need the assistance of the technical experts (the "technologists"). Nevertheless, the executive must manage the whole process and make the final call. Unfortunately, the technologists of the organization are usually not the best qualified to "pull the plug" on a failing project. It was George Pompidou who correctly said:

> There are three roads to ruin: women, gambling, and Technologists. The most pleasant is with women, the quickest is with gambling, but the surest is with Technologists.

Technologists often lead the organization astray. They have a role to play, but they are frequently motivated by technical curiosity and exclude financial considerations. Unfortunately, the typical executive has little or no content knowledge of the technology under consideration. It is common for the executive to hold their leadership position for reasons unrelated to Technology Development. Their role may have been to raise money, sell the product, or build a plant rather than develop a technology. A lack of skills can be a serious, even fatal, disconnect for the company.

When this happens, the executive **must** shore up their skill base by either learning about Technology Development (often not practical) or enlisting a trusted assistant who can add these skills. It is rarely helpful to "promote" someone from the existing group of technologists. They usually have too much ego tied up in the technology. "Outside" help is usually a surer way to do the objective work needed.

In any case, the executive must do a periodic, fundamental assessment of the technology and the connections between the technology and the target market. Furthermore, an appraisal of the Technologist group is needed. Some will be useful in helping the executive come up to speed on the key features and weaknesses of the technology. Some will be barriers to understanding. Again, the assistance of an "outside" expert can be beneficial. Once armed with a general overview, the executive **must** evaluate the connection or lack thereof between the technology and potential markets.

The executive should also periodically take stock of the resources available for the tasks at hand. The resources include money, time, and expertise. If the resources required are not available to move the technology forward, the executive needs to get additional help or kill the project. This decision is not a single "once-and-for-all" event. As the executive and the Technology Development team learn more, what resources are needed and missing will become more apparent. Hence, the executive should make frequent re-assessments and take corrective action. Again, an "outside" expert can be an indispensable source of help.

The executive has a fiduciary duty to take objective and decisive corrective action. It is not ethical to stand idly by, knowing that there is not enough "runway" for the project to get airborne. Unfortunately, executives often have difficulty aborting a project. Those with a history of this kind of failure – that is, a failure to be honest in assessments – should not be given the reigns of new projects. Investors, partners, shareholders, employees, customers, and government officials should be on the lookout for executives who function in this manner. They should remove their support from such persons and refuse to enable the fraud they commit.

It is difficult to overestimate the role of the executive in Technology Development. Unfortunately, technologists are rarely interested in or capable of making sound business decisions. Even those who have an MBA seem to have trouble ignoring the Sirens of the cool but useless gadget. Therefore, picking the right person or group to run a Technology Development project is crucial.

Example 12: Damned from the Start?

A biomass gasification start-up pursued a complex, multistep process to transform woody biomass into fuels and chemicals. Early in its development, owners hired a CEO with a background in consumer electronics to lead the organization. The company was still in the throes of several Technology Development tasks simultaneously. The CEO struggled courageously with the multiple challenges by hiring several technical experts into critical positions. A combination of personality clashes, numerous technical challenges, investor impatience, skyrocketing costs, and a few blunders proved to be too much. Owners replaced the CEO, but the die was already cast. The company struggled to right itself but finally had to give up. A promising technology landed on the ash heap of forgotten inventions.

Finally, The Technology Development Process

So far, we have looked at several aspects of technology and Technology Development – especially the early evaluation stages. Now let's turn to the nuts and bolts of a Technology Development process. A general outline is:

- Planning to learn
- Design your experiments
- Execute your plan
- Evaluation and reporting results
- Improve your experiments or wrap up your conclusions

Planning to Learn

Managing anything begins with planning – "starting with the end in mind." A surprising number of technical managers strongly believe that it is impossible to manage Technology Development. They argue that one cannot control "discovery." That is complete nonsense and often a "copout" to excuse poor management skills, general laziness, and lack of subject matter competence. All is lost if management cannot talk about "the end in mind" in specific ways. In truth, we can learn nothing of use without careful planning. The key is remembering that you are managing **processes** and not **outcomes**. Indeed, we cannot know the result of an experiment before performing it. And it is also true that many successful projects have benefited from "good luck." Nevertheless, we rarely, if ever, learn unless we have planned carefully ahead of time.

We must debunk the persistent myth that all great discoveries have come by accident. Edison popularized that idea to excuse his lack of basic scientific knowledge. It persists more recently as an excuse for many PhDs wishing to rationalize unstructured "playing" in their labs. Unexpected and even serendipitous events have accompanied many great discoveries. We could mention the events related to vulcanized rubber, Teflon, penicillin, and gravity, but nothing was learned solely by the isolated events cited in these myths. In all cases, careful experiments transformed "bizarre," unintelligible events and observations into testable hypotheses and finally into knowledge. Learning has always been a progression of events. We begin with observing and then muse, guess, hypothesize, test, and finally verify our understanding. That is the tricky business of creating knowledge.

Shortly, we will talk about the four remaining steps of Technology Development, but first, we must speak about **need** recognition. The first step is recognizing that we "don't know what we don't know." Never forget this and keep reminding yourself of this. Whereas "we don't know what we don't know," we must discern what it is we **need** to know. Until we can come to grips with what we **need** to know, we are wasting our time. Hence, all learning starts with an assessment of our needs. Surprisingly, few ever grasp this concept. Most simply will not think through what they **need** to know and then make a concerted **commitment** to fulfilling that need. This process can be both challenging and freeing. It is a challenge to the ego to begin the journey with an honest, "I don't know." Furthermore, it can be frightening to admit that there are some things that we simply **must** learn, or our cause is hopeless. Nevertheless, it is freeing in the sense that we now begin to develop criteria for action and decision-making.

It can be overwhelming to come to grips with a vast gulf between what is known and what we **must** know for success. It can be, and is often the case, that this simple exercise can end a project or business venture. All too often, the gap between what is known and what **must** be known is beyond the resources available. When this is the case, the executive should realize that they are on a fool's errand and stop the madness using the "go/no-go" decision mechanism discussed above.

Nowhere is less known about critical information than in potential markets. Frequently, the executive is forced to deal with the connection between new technology and the marketplace. As Peter Drucker often said, "The purpose of business is to create a customer." If the executive can find no connection between the technology and a market, there will be no customers and hence no business. This evaluation is not as simple as one might think. There are two fundamental ways that technologies connect to markets. These are:

- Incremental technology ("sustaining innovations") – One that improves on existing methods without changing fundamental relationships.
- Disruptive technology ("disruptive innovations") – One that is an entirely different method of solving customer needs. It results in significant changes in suppliers, distributors, and consumer habits.

Incremental technology is the mainstay of established companies. They often develop, test, and implement improvements that:

- cut costs
- improve quality
- improve customer experience and loyalty, etc.

If the technology under consideration is incremental, the executive should focus development efforts on costs, margins, quality, and more effective fulfillment of existing customer expectations. In addition, any change to the technology must have high confidence of success. Companies have too much invested in their existing market share to take risks on changes with questionable chances of success. These principles will drive the types of "improvements" developed and the criteria used to value them.

An example of an incremental improvement might be the addition of new features to a smartphone application. These features should add to the customer's overall experience and be well tested to ensure they work. If the customer does not value the changes, the development work will be a waste of time. If the new features do not work, customers may move to competitors.

If, on the other hand, the technology under consideration is disruptive, then the executive must use completely different criteria for evaluation. Disruptive technologies bring the promise of creating new markets or changing existing ones beyond recognition. Disruptive technologies often do not immediately result in lower costs, higher quality, or enhancement of existing customer experience. They do, however, bring the **promise** of one or more of these. In this case, the executive **must** be good at judging the **potential** of a technology. In addition, the executive must understand the potential **value** that customers will give to new technology. If the executive blunders this value proposition, the company will invest resources in developments that customers will not buy. Executives who do not know the value perceptions of potential customers are unqualified to lead disruptive technology efforts.

The "green industry" is full of examples where undervalued "improvements" absorbed vast resources without yielding any returns to investors. There are many examples where executives assumed that buyers would value "green" chemicals, "green" processes, and "green" products and would pay a premium for the "improvements." Unfortunately, the buyers were often far more cost-conscious than thought, and the new "improvements" did not sell. Another great example was Apple when Steve Jobs was absent. Apple could not accurately assess what its customers wanted. When Jobs returned, he made several blunders and missteps, but he excelled in guessing what customers wanted. As a result, customers forgave many problems because they got cool stuff that no one else provided.

When developing or evaluating a disruptive technology, the executive must focus the attention of the Technology Development team on those features that customers will value. If, for example, the value proposition is about ease of use, then the Technology Development team must verify that the "improvement" really delivers ease of use. Other features such as cost, reliability, and availability may also be necessary, but they do not have to be "proven" with the same level of certainty as ease of use. If customers genuinely value ease of use, they will accept some negatives to meet their value needs. In addition, they will often assume that future incremental improvements will fix some of the initial problematic features. Hence, the company developing a disruptive technology should set acceptance levels for "key features" tightly while relaxing them for less essential elements.

We can illustrate this point with handheld calculators. At first, they were expensive, bulky, unreliable, and had few functions. Nevertheless, they performed simple arithmetic very quickly and with high accuracy. They were a vast improvement over the slide rule for basic calculations. Slide rules had many functions, including logarithmic and trigonometric functions. They were cheap and very portable. They were, however, slow and incapable of handling more than three significant digits. The speed and accuracy of the handheld calculator were enough to start the displacement of slide rules. Incremental improvements came quickly. By the mid-1970s, the slide rule was replaced entirely by handheld calculators that were cheaper, faster, and had more functions. After almost 300 years of dominance in calculation aides, slide rules, and the companies that made them essentially disappeared.

Executives should heed a word of caution at this point. They cannot depend on the opinions of their technologists in this initial evaluation and setting of priorities (remember Pompidou). With few exceptions, Technologists believe that their technology is "disruptive." The inventor almost always assumes that their invention will "revolutionize" the market. This assumption is more than just an expression of invested ego. Technologists are rarely capable of assessing customer wants or needs. That is not their area of expertise. Hence, the executive must politely but firmly reject the Technologists' opinions in this matter.

> One of the most important decisions to be made in Technology Development is its positioning in the market. If the Executive fails in this regard, the whole endeavor is doomed. Therefore, owners should select Executives familiar with the target market.

Discerning customers' desires for disruptive technology is a difficult task. Early market studies are of little help. It is challenging to formulate questions that can reveal unknown, potential desires. The best approach is often to test hunches, either formally or by general discussions with "opinion leaders." The most successful examples seem to be those hunches that are proposed, tested, and refined with the direct participation of a lead executive who has been around the industry for many years.

Once the executive understands the technology and how it applies to a target market, they can assess knowledge gaps. The executive needs to determine this knowledge gap on two levels:

1. What WE know – do we know enough about the technology to assess how it can meet market expectations?
2. What THEY know – does the market know that our technology can meet their needs, or do we need to develop convincing data?

This assessment forms a basis for a development strategy. In the first case, **we** need to do more work on developing key features of the technology. In the second case, **we** need to establish enough evidence to convince buyers that our technology will meet **their** needs. Having assessed these two levels, the executive is ready to put together a targeted development strategy that focuses on real business needs.

We should expect to end this phase with more questions than when we started. They might even be more complex questions. But they should be questions that focus our attention on what we need to learn from our Technology Development program. Furthermore, we should know why they are important questions from our potential customers' perspective. Finally, they should force our Technology Development project to focus on useful (i.e., profitable) results.

Designing Your Experiments

By this stage, we should have some specific questions that, if answered, can give us confidence that we are adding demonstrable value to our technology. Assuming that these questions are nontrivial, we may find ourselves wondering how to proceed. We may have, for example, learned that the bicycle hitch we have developed has many valuable features, but it wears out too fast. This problem sounds simple enough, but what do we do next? We know how to make our bicycle hitch (which we think is cool), but we may not be experts on wear. Now what? How can we reduce wear and still have a cool bicycle hitch?

For all our desire to be objective, when confronted with "gaps" in our knowledge/experience, we must rely somewhat on our intuition – our best guesses. Later, we may be accused of being prejudiced, superstitious, silly, or crazy, but the last thing we want to do is be paralyzed or panicked. So we gather what we know, make our best guess, and then calmly and methodically act on that. We collect and evaluate more data, revise and improve our guesses, and KEEP GOING.

The Intuitive Phase

Starting a Technology Development project is like being lost in a forest and knowing that we need to find shelter before dark. Understanding the fundamental problem can be enlightening, but we are clueless without some reliable details. For example, we need to know how many hours until it gets dark, how far we are away from any potential shelter, how to build a shelter, or what tools we have at our disposal. Without some data, we cannot formulate any relevant plans.

Often, we must assess our position through some simple, easy testing, and observations. We must try a few things that may or may not help. In the case of our forest example, we might climb a tree to get our bearings. We might retrace our steps to get to a place where we knew where we were, even if we were pretty sure that was not the most direct route to where we wanted to go. We might consult a map and use our compass to get bearings from distant but visible landmarks (including the sun) to get a crude but reliable position. In the case of our hypothetical bicycle hitch, we would want to know our wear rate and how that stacks up with the competition. These things would be initial, simple testing designed to give us a general idea of our current condition.

Since we are "groping around," we want to avoid committing to expensive or dangerous courses of action until we have a clear plan that we can try and one that we can assess for progress. Random walking around kills many hikers each year and probably more businesses. The "cheapest" way to gather data is doing a literature search. Before we do anything else, we should do a thorough search for everything we can easily find on the topic. Access to the Internet has made comprehensive literature searches more practical than ever before. This step is so essential and yet so seldom followed that it bears repeating:

> Before commencing any significant Technology Development testing effort, researchers should thoroughly search existing literature to assess the current knowledge base.

There are many reasons for doing a preliminary literature search. First, we want to learn from the work of others, avoiding their mistakes and avoiding "re-inventing" the wheel. It is particularly pointless and counterproductive to spend valuable research resources on a technology owned by others. Second, there are far more ways to fail than succeed. Unless we have a plausible reason to believe that our approach might work, there is no future in just "trying stuff." Finally, never forget that random chance is always against us. This principle also bears repeating:

> There are always far more ways to fail than to succeed. So trusting dumb luck is always a bad bet.

We want to work only in those areas where the chances for success are better than just randomness. Let me illustrate by using our hypothetical bicycle hitch example.

> Suppose our bicycle hitch is "cool" because it is lightweight, can come in multiple colors, and is easily modified to hold one, two, or three bicycles by adding simple attachments. This "coolness" is all made possible because we are using anodized aluminum. But the aluminum wears poorly. It stress-hardens and cracks right at the hitch connection, dropping all the bicycles in a way to do the most damage (Murphy's Corollary).

Let's also suppose that we have three different kinds of aluminum alloy, two types of stainless steel, and one special high molybdenum alloy steel in our shop. We could use these to build seven prototypes to test. We could make these in a week and put them on seven different vehicles, each with three high-end mountain bicycles. We could put them on Colorado mountain roads to see which ones worked better. We could keep trying other metals if our first choices didn't work. Would this be a good way to proceed?

Well, the answer is, "Probably not." Experimenting at that scale will eat up a lot of money and time. It would be nearly impossible to guess which metal to use with so much unknown. The metal might not be the problem. It could be that our welding or anodizing procedures are flawed. We might also lose some cool features like lightweight or multicolors with steel alloys. A better strategy would be to search the literature for companies with similar problems. We would also want to get some outside experts to look at our "failures" to see what they think root causes might be. We would eventually like to do some prototyping and testing, but not just on the materials we have laying around the shop. We should work with others with a good chance of not getting the same stress cracks and would be nearly as lightweight as our existing aluminum alloy.

This approach is the kind of thinking and "library" work we need to reduce the scope of our inquiry from a myriad of random possibilities to a few more manageable and plausible options. Of course, we would probably do more than what we describe here, but the idea is to do enough to have a reasonable plan before we start doing physical testing. We will, of course, have several options. One might be a change in metals. Another might be a change in design that changed weight distribution or replaced a single component. Which of these will be best? Which should we try first? Do we need to try them all?

Again, we apply our intuition and maybe some "expert" opinion. Must we build a complete bicycle hitch and load it with $18,000 worth of high-end mountain bicycles to do a comparative test? What if we make a little bench test unit that loaded the critical failing parts with equal loads and compared the failure rates of our prototypes and our current model. This device may not tell us that our change will "beat the competition," but it could tell us that we probably made a measurable improvement to our current model, which could be helpful. It could also tell us which changes give us the biggest bang for the buck.

We would continue this thinking process with other factors of interest. In this initial phase, we must remain open-minded and vigilant. We need to be looking for the unexpected and document that carefully. We aim to:

- "discover" the most important factors (variables we control),
- develop key hypotheses about the relationships between factors and results (outcomes of interest),
- hone our abilities to control the experiment and measure what seems to be necessary,
- develop a "history" to serve as a guide when we must estimate the connections between factors and results, and
- keep costs down as we "grope around."

Some researchers have told me that they "never" run an experiment without a complete experimental design ("design of experiments (DoE)"). They indignantly proclaim that they have no faith in intuition or common sense. These researchers rarely discover anything and are experts only in wasting resources while chasing mirages. Furthermore, they confuse themselves and their organizations with unprofitable complexities. They are the ones who starve to death in the woods a half-mile away from a McDonald's, convinced that the yellow glow in the sky to their south must be the northern lights.

The executive must allow a certain amount of unstructured but well-documented guessing and probing. There needs to be some "informal" data gathering to make reasonable guesses about what is going on. At the same time, the executive must be pushing for continuous progress toward the next phase. That phase is a formal deductive phase, where the development team will formally state and test specific hypotheses with planned experiments.

The Deductive Phase

In the intuitive phase, we "learned" a lot, but our confidence in what we have learned is still unknown. We based our probing on our initial "understanding" of the technology. We made many assumptions about how it worked and how well we could measure how well it worked. As a result, we were very biased in our approach. We have not yet developed objective evidence that supports our general claims and establishes our confidence in those claims. We need to know that our claims are objective, unbiased estimates of the expected outcomes – that these are not just our opinions supported by insufficient data. We also need to have objective information indicating the variability around that expected outcome.

> There is a grave danger at this point. We may have collected just enough information to give us unwarranted, unmeasured confidence in our original prejudices. However, we must not stop here; we must think things through. We must gather enough information to support our confidence in our claims objectively.

This last point is especially crucial. Not only do we need enough objective evidence to support our claim, but we also need enough to supply an objective confidence level to skeptics and detractors. For example, we might have evidence supporting a claim that the return on an investment in a random stock on the New York Stock Exchange is around 10%. This isolated information is not very useful until we know how much variability there is in that average return. We certainly should not expect to buy stocks randomly and get a 10% return on every stock purchased.

So how do we do this? Let's first speak in generalities, then get into the formal logic, and then dig into some of the mathematics behind our formal, objective approach. But, first, let's think about what was "wrong" with our intuitive approach. In this, we must always remain the "skeptic" if we want objective evidence that we will use to silence critics and gain the support of third-party investors and stakeholders. We will stick with our bicycle hitch example to illustrate points.

As you might expect, we need to define terms more formally. Therefore, we will use the following terms frequently:

Factors: These are **variables** that we will manipulate as part of our testing program. They are the variables that we think may affect the outcomes of our testing. So we deliberately alter these to show how they influence results.

Results: These are the **outcomes** of interest to us. There are usually many such outcomes, though we may focus on just a few.

Note: You will see factors/variables and results/outcomes used interchangeably.

During our intuitive steps, we gathered data based on a review of literature, the availability of historical data, and some preliminary testing/trials. We assumed many things about factors and results that we thought were reasonable and important. We did not test every possibility or try every possible combination of factors. One of the things we will have to do is give evidence that the factors we looked at made a significant impact on the outcomes we think are essential. For example, in the case of our bicycle hitch example, we need to show that changing the material, design, or construction methods has a significant impact on "wear."

We probably made at least two subtle assumptions that we should review. First, throughout the intuitive phase, we assumed that we could (1) control all the relevant factors and (2) we could measure results effectively enough to determine "good" results from "bad" results. We need to test these before moving on to the deductive phase of our work. Part of our deductive work will include generating objective evidence that we are on the right track with the controllability and relevance of factors, along with our ability to differentiate "good" from "bad" results.

We first need to determine how confident we are that the factors we are testing are relevant. If we overlook something here, the conclusions that we draw later could be completely irrelevant. It is essential to look skeptically at any "historical" data we found and used rather than generated ourselves. We need to investigate the quality and circumstances of data collection, looking for uncontrolled or confounding conditions.

We also need to clarify terms regarding the factors and results under study. When we are informal, we tend to be "fuzzy" with our words. In our bicycle hitch example, we assumed that "wear" was the critical result of interest. We thought that "wear" was related to design, materials, or manufacturing processes. We are probably not looking at "wear" but rather "failure rate." The hitch was not slowly wearing out but failing catastrophically. Failure rates are much different from gradual changes and governed by other statistical models. We need to be clear on what we mean and proceed accordingly.

We also assumed that the failures had serious sales consequences. That's probably true, but how serious? Furthermore, what is the relationship between mean time to failure and sales? We may find that we need to develop "measuring systems" that have nothing to do with design, materials, or manufacturing processes but "measure" customer expectations about mean time to failure. We might find that it is not a hitch failure problem but rather a customer use, training, or expectation issue.

We should not leave the topic of factor relevance without revisiting historical data. Historical data is also called "secondary data." Others collected these data without designing the collection methods for our specific purpose. When we collect data where we have controlled the collection, the data are called "primary data." The most significant difference is that we try to control all factors relevant to us. Furthermore, we do random sampling and replicating of trials so that factors we cannot control "average out." Although secondary data are cheap and plentiful and frequently used by "data scientists" in "data mining," the data could be utterly useless for our project because the researchers did not control a factor crucial to us.

Let's go back to our bicycle hitch example. Suppose we had turned three engineers loose during the intuitive phase and let them try a few things to see what they could do to fix our problem. Then, to keep costs down, we gave them only a week to "play around" with the problem:

> The first engineer built a rack out of stainless steel instead of aluminum. He then drove over Engineer's Pass with a full load of mountain bikes without incident. Another engineer added a final heat-treating step to the current aluminum design. That engineer designed a bench-scale fatigue tester and used it to compare the time to fatigue cracking to that of the standard design. The new heat-treating method showed an increase of time to fatigue of 50% compared to the standard procedure. The third engineer used the second engineer's fatigue tester and tried making the hitch out of a different aluminum alloy with the standard production line. This final engineer found the change made fatigue cracking 10% better. Would you be ready to switch the production line over to stainless steel, adopt a new heat-treating method, or abandon looking for other aluminum alloys?

The above would be a pretty typical intuitive approach. We did "learn" some things, but we didn't gain much confidence in any of them. We don't know if one trip over Engineer's Pass proves that stainless steel is the best choice. We don't know that the fatigue measuring device works at all for measuring failure rates or what kind of improvement we need. Finally, we don't know if the different alloys "solved" the problem or not. We have several hypotheses to be tested and one new candidate measurement technique. However, a week of "playing around" has yielded several possibilities.

We would be way ahead if we validated the bench-scale fatigue tester and found that it gave valuable and reliable results much faster and cheaper than hauling around expensive mountain bikes. This device could be the breakthrough that gives us a competitive edge in our ability to measure "failures" quickly and cheaply. Furthermore, this one week gave us some hypotheses to test (e.g., stainless steel, heat treatment, a different alloy). It should have also given us additional, related ideas to investigate.

It is precisely at this point that things can go very wrong. We have invested time, money, and egos into solving the problem, but are we done? We have some clever ideas, a little bit of data, and probably three champions for three different solutions. The executive must step in and insist that the development team goes from "playing around" to the serious business of proving performance. We will try to describe this transition with the hope that your Technology Development team and management organization can make this transition successfully. It is important to note that we write this for industrial research, emphasizing making money and conserving time and resources.

In the deductive phase, the Development Team must state hypotheses in measurable terms that lead to statistically based evaluation. Then, they must design experiments to test those hypotheses and measure relationships between factors. In the case of industrial research (i.e., R&D for profit), the goal is to obtain a statistically valid model that can do cost/benefit analysis for business purposes. A model is still necessary for scientific or philanthropic research (i.e., the goal is something other than profit), but the evaluation criteria will differ. For example, the model may focus on elucidating mechanisms or searching for new physical laws.

The transition from the intuitive to the deductive phase is usually messy and awkward. Very little is really "known," although much is suspected. Opinions abound. As a result, it can be difficult to make objective decisions about proceeding. Nevertheless, the development team **must** make decisions – sometimes arbitrarily. Usually, the intuitive phase generates interest in more factors (variables) than can be managed in a testing routine. The quasi-formal testing of the intuitive phase often reveals unexpected variables and puzzling results.

In many cases, even the number of relevant results has grown. It is often the case that the desired result (e.g., yield) turns out to be a combination of two or more independent factors complicating the entire development program. The development team must set the testing priorities. These priorities will include:
- Which ideas to test first?
- How many ideas to work on?
- Who will work on which ideas?

During testing (or "trials"), the development team will set factors at varying levels (two or more) and measure results. All other factors (variables) are held as constant as practical. Each set of factors and results is a "run." Rarely can the development team vary more than a handful of factors (4 to 6) and roughly half as many results (2 or 3) simultaneously. The required number of runs grows with more factors and results. The number of runs (dozens to hundreds) frequently becomes too large for management to handle in the operational or budgetary cycle. Even in academic and philanthropic circles, the time and resources needed to manage programs requiring hundreds of independent trials are rarely available. Hence, it is often the case that a strategic decision becomes necessary. The Technology Development team must reduce the number of factors and results, thereby reducing the number of independent tests.

Example 13: Operating a Pilot Plant Simulating a Mineral Extraction

A research company built a pilot plant to test the yield of oil extracted from rock for their patented process. The size of the pilot plant allowed several tons per hour to be processed. Operating the pilot plant required around-the-clock crews of a minimum of several well-trained operators and support personnel. The pilot plant was aging and had undergone multiple repairs and modifications. As a result, the plant became unreliable and often shut down for reasons unrelated to the testing. In addition, there were several "competing" theories about which factors were the most important to achieve desired results leading to long and complicated trials. All these conditions affected operational control and led to confounded, unreliable results. Only after the Technology Development team implemented extensive repairs and instituted strictly controlled and much less ambitious research plans was valuable data available from the pilot plant trials.

Technology Development team must make many tough decisions setting priorities for the development plan. They must prioritize testing those factors and hypotheses that significantly impact the most important results. Here, the executive must take a leadership role in forcing prioritization. The executive can serve as an indispensable referee and guide to sorting priorities by bringing attention to potential impacts on organizational costs, customer benefits, and investor/stakeholder requirements. Some factors and results will have a minor impact on organizational goals. These can be "shelved" until later.

The Technology Development team takes these decisions knowing they risk overlooking important factors or results. Missing important factors can prejudice and confound later choices. Therefore, the Technology Development team will frequently opt for a "screening" strategy early in the development process. A screening strategy is a simplified testing program that focuses on many factors with minimal variation of each factor. The idea is to find the few factors with the most significant impact on results. Then, the team can drop those factors with little impact results from further testing. Screening can be a helpful strategy to refocus the research efforts on key factors for detailed further study.

> When the number of possible factors is large, a screening strategy is often the best way to help us focus on the few, most important factors.

To this point, we have been talking about factors and results as if they were independent objects that we wish to describe. That is hardly the case. Factors are essential only in the way they impact results. Therefore, the relationship(s) between factors and results is of most interest to us. We state these relationships in the form of hypotheses – often mathematical formulas. Hence, we express how we think factors affect results in quantifiable terms to measure and test. If we can confirm these hypotheses, we will have a mathematical model for how our technology works. We can use that model to predict financial performance, validate our business model, and "prove" the value of our technology.

Unfortunately, this means that we must do some mathematics – specifically statistics. Don't be alarmed at this point. We will try to explain this in simple, intuitive terms and avoid complex mathematical formulas. What we are trying to do is to get a "feel" for how sure we are of what we think we know and put it into a number (a "statistic") that puts that "feel" into an objective term. Hang in there. All people working in Technology Development must understand the basics of assessing uncertainty. This basic understanding of uncertainty is often called "statistical inference." Without a basic knowledge of statistical inference, it is impossible to compare alternatives objectively. We are not seeking certainty, just the ability to make "good guesses" versus making "stupid blunders."

Making Sense of Uncertainty

The best way to start thinking about statistics is to use simple examples that don't require complex mathematics. Let's start with throwing a single die (i.e., "one dice"). The outcome is not entirely predictable, but we know we will not get a 0, 7, or 3.5. Instead, we will get a 1, 2, 3, 4, 5, or 6. We also know that if the die is "fair," we cannot predict which of these we will get in a single throw, but they will all be equally probable. Simple math tells us that one-sixth of the time, we will get a 4. Similar arithmetic tells us that one-third of the time, we will get a 4 or a 5. If we were to roll a die hundreds of times and record the results, we would find:

1. after a few rolls, our results would not be exactly what we calculated, but
2. after many rolls, our results would get closer and closer to the one-sixth and one-third that we calculated.

Example 14: Playing with Statistics

The best statistics course I ever took was in the summer of 1964. That summer, my two brothers, two neighborhood kids, and I re-played the entire 1963 Major League Baseball season using a statistics-based game called Strat-O-Matic Baseball. We divided up the teams among the five of us. We became the managers and expected our managerial skills to have at least some impact on the results.

The game uses three dice, one red and two white, which are thrown to determine the outcome of an at-bat. The possible outcomes are determined by events (like a "strikeout" or a "home run") listed on two cards – one for the batter and the other for the pitcher. The red die determines which column (1 through 6) the event will be from, and the white dice indicates which row (2 through 12). Columns 1 through 3 are on the hitter's card, and columns 4 through 6 are on the pitcher's card. Hence, the possible outcomes reflect the combined skills of the hitter and the pitcher. Good hitters increase the chances of a hit or walk. Good pitchers increase the chances of an out.

We were baseball fanatics. We saved all the box scores from every game (several ring binders). We compiled team standings and individual key statistics on most of the players. We even had an All-Star Game using the players with the best performance from "our season." The All-Star Game, by the way, resulted in several arguments over who should be an All-Star. In "our season," Marty Keough of my Cincinnati Reds was batting well over .300 at the All-Star break. Marty's 1963 batting average was an anemic .227. I insisted that Marty should be an All-Star. I won that battle. Nevertheless, when Dick McAuliffe of the neighborhood kid's Detroit Tigers was outperforming Bobby Richardson of my older brother's beloved Yankees, the rules had to change. Suddenly Richardson's 1963 .265 batting average trumped McAuliffe's 1963 .262 batting average!

At first, our "managerial skills" seemed to matter. The boys that took the game more seriously and tried hard to get lineups "right" seemed to be doing better. As the "season" wore on, and we all got better at the game's mechanics, the results of "our season" and the 1963 season began to converge. By the end of "our season," Keough had faded into oblivion, Richardson and McAuliffe both hit in the .260's, and my wonderful Dodgers beat those pesky Yankees. Not in a 4–0 sweep as in 1963, but in a 4–1 rout.

That summer, I learned that you could beat the odds in the short term, but they will catch up to you if you play any game long enough.

So now, let's play some mind games based on rolling a single die. Consider the following games:

Game 1: I ask you to bet $1 on the number 4. I agree to pay you $2 (your bet plus $1) if the number 4 comes up on one roll of the die. Will you play?

Game 2: I ask you to bet $1,000 on the numbers 4 and 5. I agree to pay you $5,000 (your bet plus $4,000) if a 4 or a 5 comes up on one roll of the die. Will you play this game?

Game 3: I ask you to bet $1,000 on the number 4. I will pay you $2,000 (your bet plus $1,000) if we roll the die 50 times and the number 4 comes up 8 or more times. Will you play that game?

Unless you were extremely desperate to get an extra $1, you would not play Game 1. You should expect to win only one-sixth of the time. Hence, your "expected winnings" would be $2 × 1/6 = $0.33. You might win on the first roll, but if you played this game for a long time, I would win 5 times more often than you. I would pay you $2 once and a while, but I would win your $1 five times more often. It would be a blunder for you to commit to playing this game, and the more you played it, the poorer you would become.

Game 2 would be a much better game for you. There is a chance you would lose your bet of $1,000, but the return is 5 times that. So your "expected return" would be $5,000 × 1/3 = $1,666.67 (remember that a 4 or a 5 is a win for you). In general, you would want to play this game, and the more you play it, the better you would like it.

Game 3 is quite a bit more complicated. We know that the "expected winning" for a single throw is $2,000 × 1/6 = $333.33. But we are rolling the die 50 times. So we need to figure out the chances that we would get 8 or more 4's in 50 rolls. We can calculate an exact solution, but let's make a quick guess first.

We know that if we throw the die "a lot," we would get the number 4 one-sixth of the time. It turns out that 50 independent tests are "a lot" in statistics. Hence, you would expect to get the number 4 close to 50 × 1/6 = 8.3 times in 50 rolls. You, of course, would never get exactly 8.3 in any single set of 50 rolls, but if you played this game many times, the average times you got a 4 would close in on that number. Since you would win the bet if the number 4 shows up 8 or more times, and you expect it to average MORE than 8 times with 50 throws, you should suspect that this might be a "good bet" for you.

There is a pretty simple way to precisely make this calculation using the binomial distribution. It works almost any time there are a countable number of trials or tests that result in just two outcomes, such as a "win" or a "loss." You would "lose" your bet only if the number 4 came up 0, 1, 2, 3, 4, 5, 6, or 7 times in 50 roles. Otherwise, you would win your bet. Using the binomial distribution, we can plot the probabilities of getting the number 4 precisely 0, 1, 2, 3, 4, 5, 6, or 7 times in 50 roles. If we sum these, we get the probability of you losing. We can state this as a percentage. The likelihood of you winning is simply 100% minus the percent probability of you losing.

In this case, the chances of getting 0 to 7 occurrences of the number 4 are about 0.3911 or 39.11%. This probability is the probability of you losing the bet. The probability of winning your bet is 100% – 39.11% = 60.89%. So now, are you willing to risk $1,000 to get an expected return of $2,000 × .6089 = $1,217.80? That is about a 22% gain. That's not too bad, but maybe not enough for you. And this is where statistics ends, and other factors come into play. Statistics would tell you to take that bet if money is all that matters.

Russian Roulette is another game with probabilities like rolling a die – 1 in 6. Nevertheless, the "costs" of losing at Russian Roulette are so dire that no one in their right mind would evaluate it strictly on monetary returns. Although this is an extreme case, nonmonetary factors always come into play. As a rule:

> Don't make a habit of "betting the farm" on a single "roll of the dice." Instead, work the odds in your favor, manage your exposure so that you can stay in the game, and reap the benefits over time.

Game 3 points the way toward a more sophisticated analysis of uncertainty. If we look at a graph of the probable outcomes from that game, we will notice something very interesting (see Figure 1).

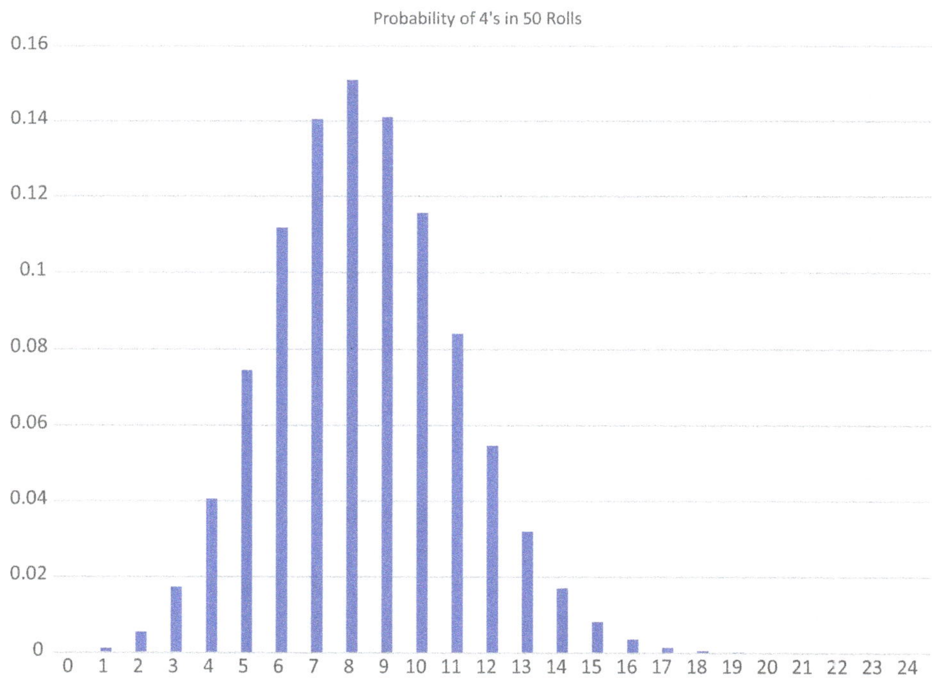

Figure 1: The probability of rolling a "4" in 50 rolls of a die.

Notice that the graph of probabilities is beginning to look like the "bell curve" of a normal distribution. This is no accident. Whenever the number of possible outcomes grows large, the "distribution" of those outcomes usually approaches a bell-shaped curve. Without going into the theory of why, we will simply state that whenever measurements are being made with many possible values (whether many counts or "continuous" measurement), we will usually be looking at bell-shaped curves of probabilities that can be modeled using normal distributions. In most cases, we will be basing our decisions on normal distribution models. This distribution usually remains true when many random factors come into play simultaneously. Most random factors are normally distributed. Combining these factors usually preserves the normal distribution. Thus, we can make models for uncertain events based on the normal distribution. We will use those models to help us make favorable bets and avoid making big blunders.

Simple Statistics – Population, Sample, Average, and Standard Deviation

We can't get very far with "thinking in bets" until we do some math. So now is an excellent time to start with some simple ideas – population, sample, average, and standard deviation.

The average is familiar to us when doing measurements. We make measurements, add up the numbers, and divide by the number of times we make a measurement. That's our average. The number we get is called a "statistic." It is a number that represents something about the population. In this case, it is the statistic called the "average." This same "statistic" is also called the "mean" and the "expected value."

> The term "expected value" can be a bit problematic. However, it makes perfect sense when dealing with a normal distribution because the "expected value" is also the most frequent. With other distributions, this may not be the case.

We often use the average to represent the entire group of interest. For example, we might want to estimate the number of apples we would receive if we bought 20 pounds of apples out of a farmer's truck. We could take a few apples, weigh each one, and divide one apple's average weight into 20 pounds. The answer would be an estimate of the number we would get in 20 pounds. To be more precise, we took a sample of apples from the farmer's truck and used the average weight per sampled apple to represent the average weight of the apples in the truck. The group of apples in the farmer's truck is the population of apples from which we took the sample. The equation for the average is

$$\mu = \sum_{i=1}^{n} \frac{r_i}{n} \tag{2}$$

where μ is the average, r_i is the ith result, and n is the number of results (or samples).

If the number of apples in 20 pounds were critical, we would want to know more than just the average. We would like to see the variability or scatter to expect. The standard deviation is a measure of variability that is very useful. It is calculated by first taking the weight difference of each apple from the mean, squaring that difference, and taking the average of those squared differences. The square root of that average value is the standard deviation. The larger the standard deviation, the more scatter in the data and the more uncertain we will be that the number of apples in 20 pounds will be close to our estimated value:

$$\sigma = \sqrt{\sum_{i=1}^{n} \frac{(r_i - \mu)^2}{n}} \tag{3}$$

where μ is the average, r_i is the ith result, n is the number of results (or samples).

Let's illustrate these ideas with another relatively simple example. Let's suppose that we have done some studies and found that our average cost to produce a commodity product is \$100 each. Let's also assume that costs follow a normal distribution with a standard deviation of \$5. Finally, let's also suppose that we have good information about expected commodity prices. They also seem to follow a normal distribution with an average of around \$135 and a standard deviation of \$15. We show these distributions in Figure 2.

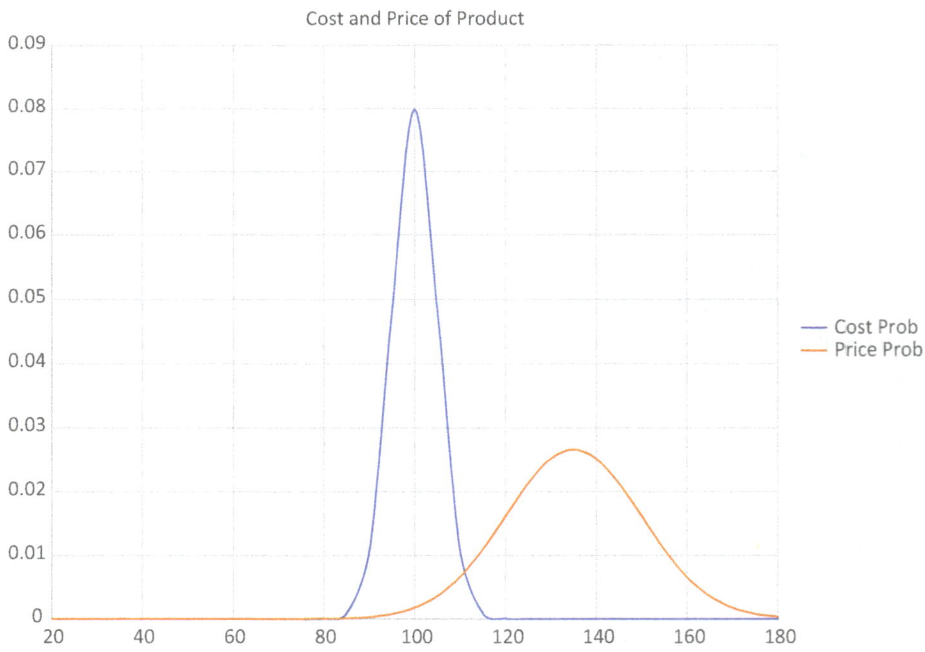

Figure 2: Example price/cost distributions.

We can see a reasonable chance that prices would be below costs at times. If we could not delay selling our product, we would undoubtedly want to know how frequently we would be losing money. Since these are both normally distributed, their difference (i.e., the profit) would also be normally distributed and easy to calculate. The expected average profit would be the difference between the mean of the expected price and the mean of the expected cost (i.e., $35). The standard deviation of the profit would be the square root of the sum of the squares to the two distributions. The expected profit distribution would look like Figure 3.

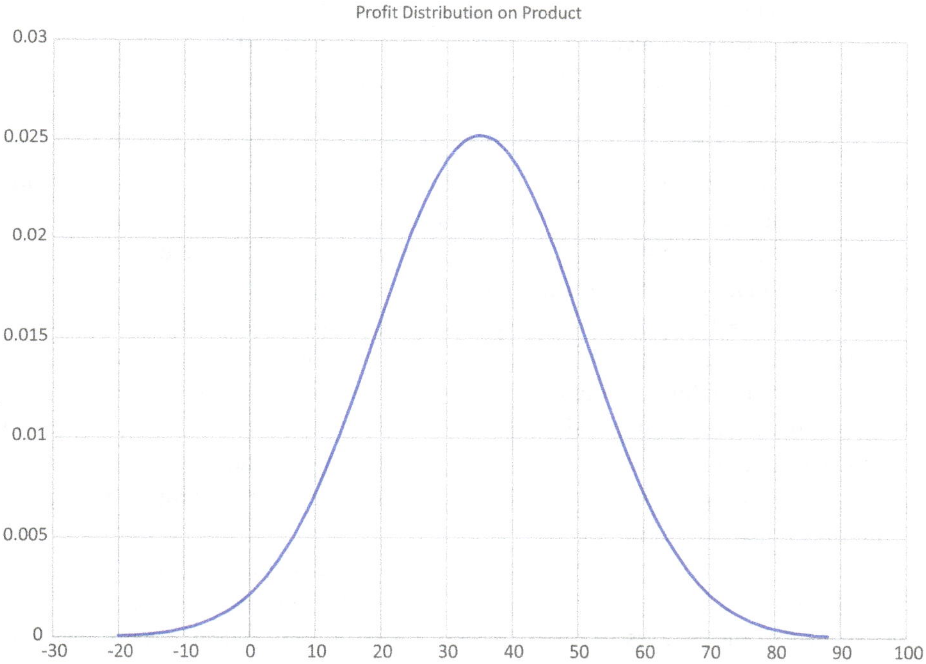

Figure 3: Example profit Distribution.

As shown in Figure 3, we would not expect the profit from this operation to be constant. It should average $35 per product but could be as high as $80 under the best of circumstances and as low as −$20 per product if costs are running high when the product must be sold into a down market. We can also predict the probabilities of these excursions and plan for them. We could consider building storage for the product inventory (if practical). We could negotiate longer term, fixed-price contracts with some major customers. This option would trade off some of the profit during high prices for protection against losses when prices are low. If there is an established commodity futures market and we had some idea of when prices would be low, we could even consider selling a forward or futures contract on the product to protect against losses caused by low prices.

The above examples are designed to get you thinking in terms of bets. This same thinking will be needed when managing Technology Development. There is, however, a bit of twist. In Technology Development, we often decide if a claim about the efficacy of changing a process, procedure, or product is true. We usually compare the status quo with a proposed change when developing technology. The hope is that the proposed change will improve results far more than the proposed change's cost, danger, or inconvenience. Hence, we need to take careful measurements of the input factors and the results. We need to have enough confidence in those measurements to make objective decisions and predictions.

In Technology Development programs, we frequently compare two or more data sets and ask, "Are these two sets of data the same or different?" We often compare results from the standard process (or status quo) with our (hopefully) "improved" process. We do this by looking at results from using the "standard method" and the "improved method." For example, we might compare popping popcorn using the "standard" setting on the microwave to setting the microwave at a specific power for a particular time and counting the total number of unburned, popped kernels. We will see "scatter" in the numbers that may obscure any real difference or give us a false impression about how good our "improvement" really was.

Let's illustrate this using a concrete "mind experiment." We will assume that we are popping corn in a microwave using the "standard" and an "improved" method. We will count the unburned, popped kernels. We won't actually pop the corn, but we will generate data using an Excel random number generator. We will assume that we will pop seven different bags of microwave popcorn by each method. We will set the means and standard deviations in the Excel random number generator to generate the normally distributed results.

Two sets of seven randomly generated points are shown in Figures 4 and 5. In both Figures 4 and 5, the averages for the "standard" set and the "improved" set were set at 100 and 110, respectively. In Figure 4, the standard deviations were set at 10. The standard deviation is, of course, a measure of the scatter in the data. The higher the standard deviation, the higher the scatter. In Figure 5, the standard deviations were set at 5. You should notice that both means of the data differ by about 10.

It is easy to see in Figure 5 that the improved method is better than the standard method. Figure 4 is not nearly as obvious even though the two sets have the same average difference. We compare data sets like these using the "*t*-test." Using that test for the actual data in Figure 5, we conclude that the improved method is "better" with 95% confidence. Using the *t*-test with Figure 4, we would NOT conclude that the improved method is better. The data would not prove they are different at a *95% confidence limit*. Because of the scatter, we must conclude that the improved method data is NOT different from the standard method data; therefore, there was no improvement. This conclusion is what is meant by statistical inference. We make decisions based on what the statistics tell us rather than a "hunch" or bias.

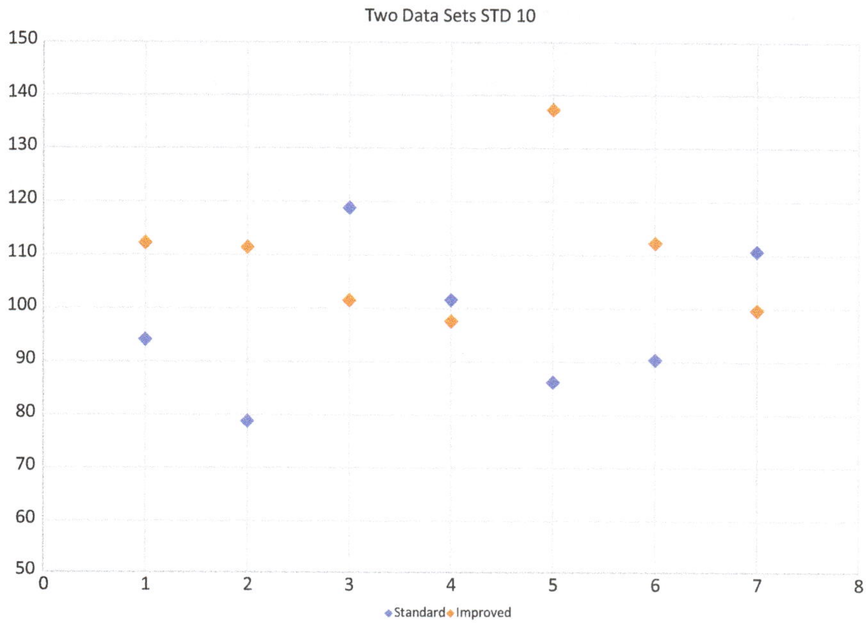

Figure 4: Two example data sets with seven SXs and STD = 10.

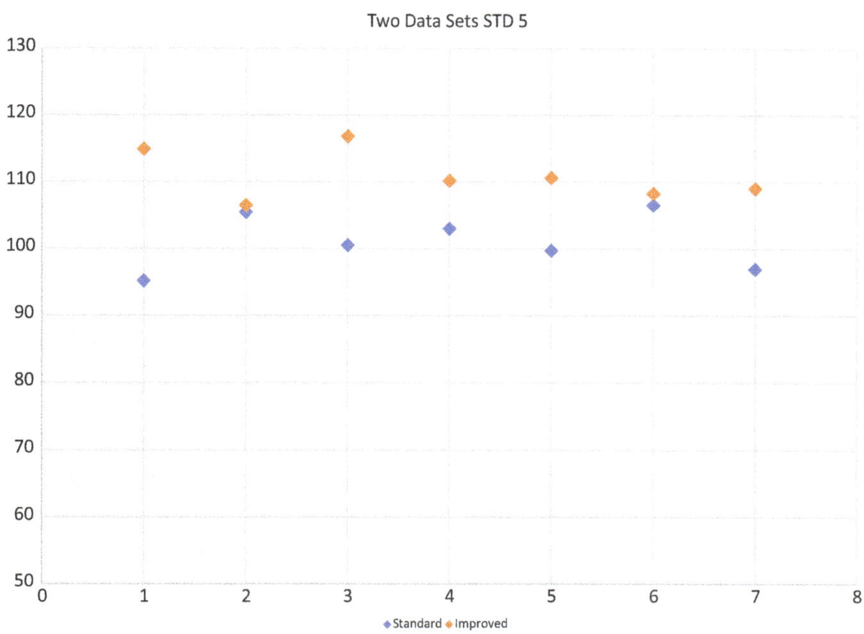

Figure 5: Two Example Data Sets with 7 SX's and STD = 5.

These are randomly synthesized data points to illustrate a point. We "know" the methods are different because we made them so, but the scatter in the data is too high to prove that they are different. However, had this been actual data, we would "know" nothing but what the data told us. For data in Figure 4, we would have come to the wrong conclusion – the improved method data were no different (i.e., no "better") than the standard method data. More testing might help us conclude that the improvement is real, but the seven trials would not be enough. So, let's try this in Figure 6. We will use 30 data points:

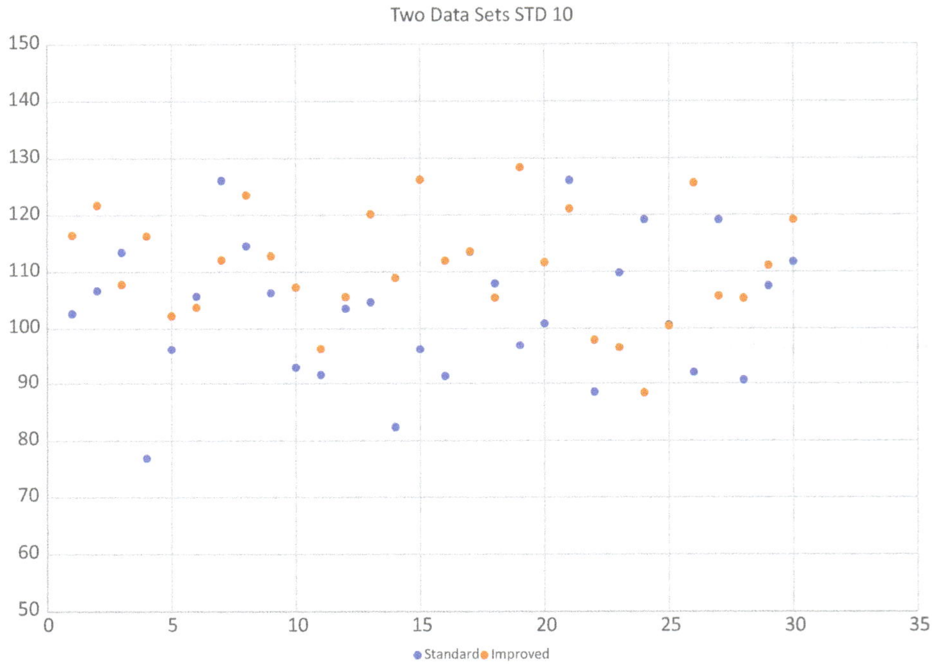

Figure 6: Two example data sets with 30 SXs and STD 10.

It is easier to see if we use a pivot table to create a histogram. We put the results in "buckets" and count the number of results that go in that "bucket." We show this in Figure 7.

Now we can see more clearly that the "improved" method appears "better" than the "standard" method. Furthermore, applying the *t*-test for two means with these two sets of 30 samples tells us that the averages are different at more than a 95% confidence. Hence, as we would expect, more test runs help us discern smaller differences in means when the scatter is high.

In this process, we have illustrated the "power" of our testing program to discern differences in the standard and improved methods. If a difference of 10 (i.e., 110–100) were an important (as in "profitable") difference, we would want to run

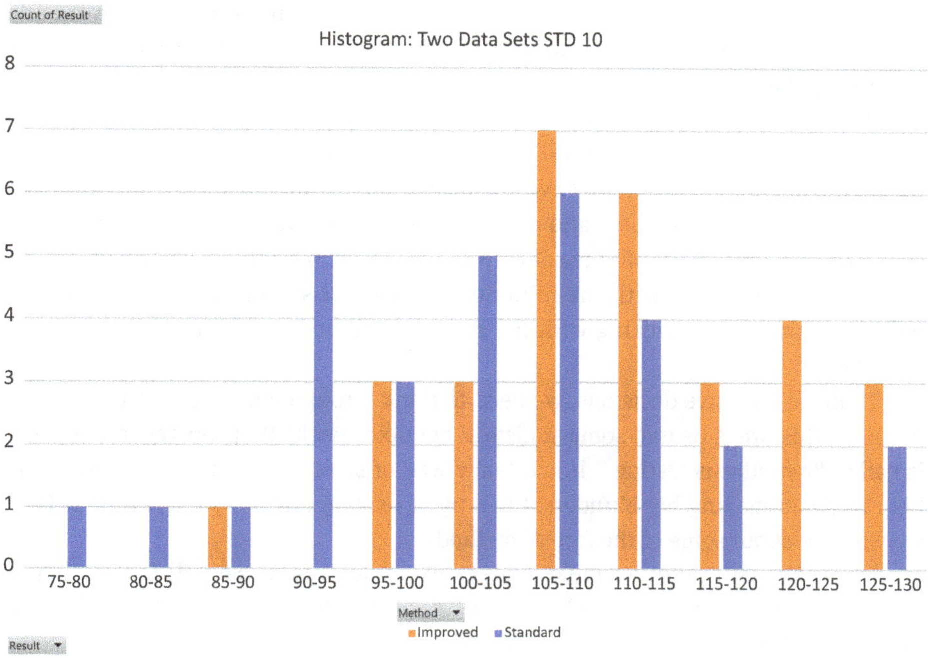

Figure 7: Histogram of example data set of Figure 6.

more than seven samples to see if the difference was 10 or more with a high (>95%) probability. If it took more samples to "see" that difference, we would want to run more or change what we were doing to have more "power" to discern a significant difference.

We try to do this "power" evaluation upfront when designing a testing program. We look at our ability to reproduce results (i.e., the standard deviation testing for results) and the minimum amount of a difference (i.e., "improvement") that would be of value to us. We then estimate the number of sample tests we would need to have confidence that we could "detect" that minimum difference. This planning process should give us confidence that we will have conclusive rather than inconclusive data after our testing. If we can conclude that the "improvements" didn't work, that is not a complete failure. We can abandon that idea and move on to another one. If we have inconclusive data, we have wasted our time. That is indeed a failure.

This approach is known as a power analysis. It is a standard feature in any good Technology Development program and one of the first steps in going from the intuitive to the deductive phases. During the later stages of the intuitive phase, we carefully test our ability to reproduce results under controlled testing conditions. We don't need perfect data, but it does need to be relevant and reliable. We also do "mind experiments" on what changes in results will be economically significant. These form the basis of our power analysis. From there, we use our DoE to design a sampling and analysis plan that should give us relevant results.

Unfortunately, 30 test runs, even in a laboratory setting, need much work and cost. Thirty test runs could be prohibitively expensive, especially if requiring pilot plant data. Each pilot plant run could cost hundreds of thousands of dollars. We would stop and think about what is causing the data scatter and reduce it if practical. Changing how we run the tests could significantly reduce the number of tests we need. We should think about every aspect of our testing, including our sampling methods, analytical methods, and ability to control the processes more thoroughly. Once we had done everything feasible to reduce random scatter in our practices, we should run the testing with the minimum number of samples we thought would help us decide about sticking with the standard method or adopting the improved method.

To make objective decisions, we need to think through our logic carefully. Unfortunately, this drags us into some statistical "double-speak" that can get confusing. It is called "hypothesis testing." It is a fancy way of structuring logical scenarios that help us avoid making big blunders. I will try to explain it in simple terms first. Then we will talk about some of the formal methods.

Let's start with another make-believe example. Again, we will use statistical software to generate "samples" that we "know" follow the normal distribution. Then we will treat them as if they were experimental observations and use them as data points to illustrate how to interpret results. Let's suppose that we have three different processes we want to test. We will call them Process A, Process B, and Process C. They could be three painting techniques, metal stamping procedures, or drug dosages we want to compare. We will use our statistical software to create three sets of "sample results" with averages of 100, 105, and 110, all with standard deviations near 10. We will test two different scenarios, one with seven samples for each process and one with thirty samples.

Our goal is to determine if one or more processes are "better." Unfortunately, the scatter in the data makes this more difficult. Intuitively, we realize that running more samples should give us more confidence, but how do we consistently deal with all these data and interpret results? Analysis of Variance (ANOVA) is a powerful technique to compare many different experiments. The method compares the standards deviations with the differences between the average results and returns several numbers that reflect the significance of those differences. Most good software packages provide graphing functions to visualize these differences.

In Figure 8 (calculated in DesignExpert® by StatEase), we plot the data from the seven sample "experiment." DesignExpert® plotting function works well on a computer screen but does not print well. Hence, I have replotted the results in Excel (with considerable pain). The small red 'x's' indicate individual data points. The yellow columns indicate the averages for the processes. The green lines with bars at the ends indicate the "Confidence Interval." A Confidence Interval estimates the range of the average. For example, we are 95% confident that the average for

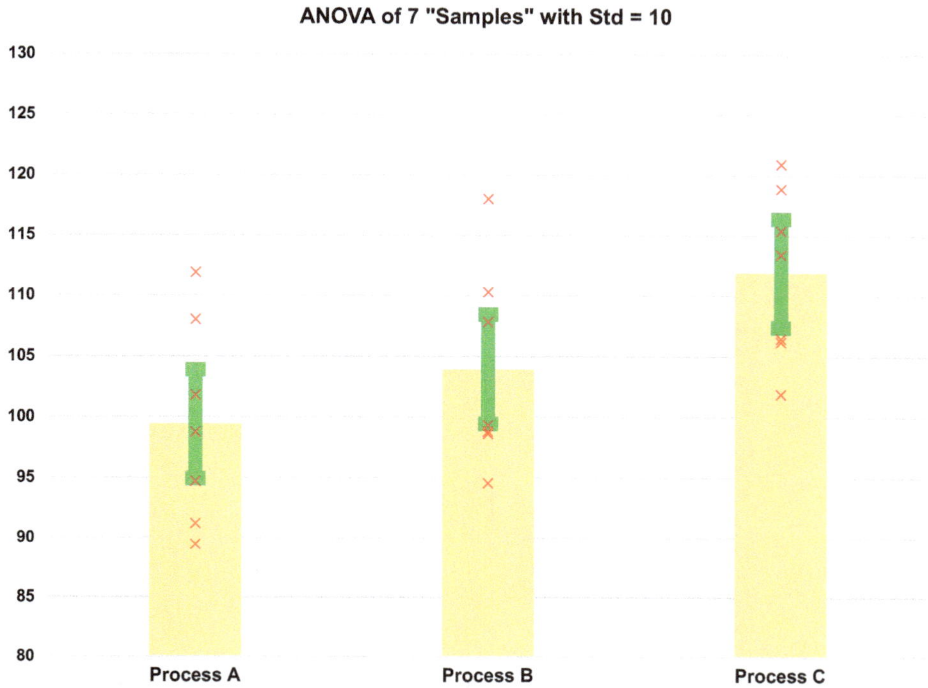

Figure 8: "CI Plot" of Example Data with Seven "Samples".

Process A is between 95 and 104, Process B is between 99 and 108, and Process C is between 107 and 116. Notice the overlap between the intervals. We can see that Process C is better than Process A, but we are not very sure about Process B compared with the others.

If we had to decide which method to use based solely on these data, we would pick Process C over Process A, but we might be left wondering about Process B – especially if there were other considerations. For example, if the cost of Process C were high compared to Process B, we would wonder if this was a good bet. So we might wait until we can gather more data. We could, for example, run 30 samples instead of seven. We know that changes the statistics dramatically from our t-test done earlier. Figure 9 shows us the "CI Plot" with 30 samples. We see no overlap of the 95% confidence limits for the averages. Hence, the differences in the averages are statistically significant (which, of course, we already "knew" because we "manufactured" the data).

We should notice three things:

1. The Confidence Interval is around the average and NOT individual results,
2. More samples make us more confident of the AVERAGE, and
3. Larger differences between the averages and smaller standard deviations require fewer samples.

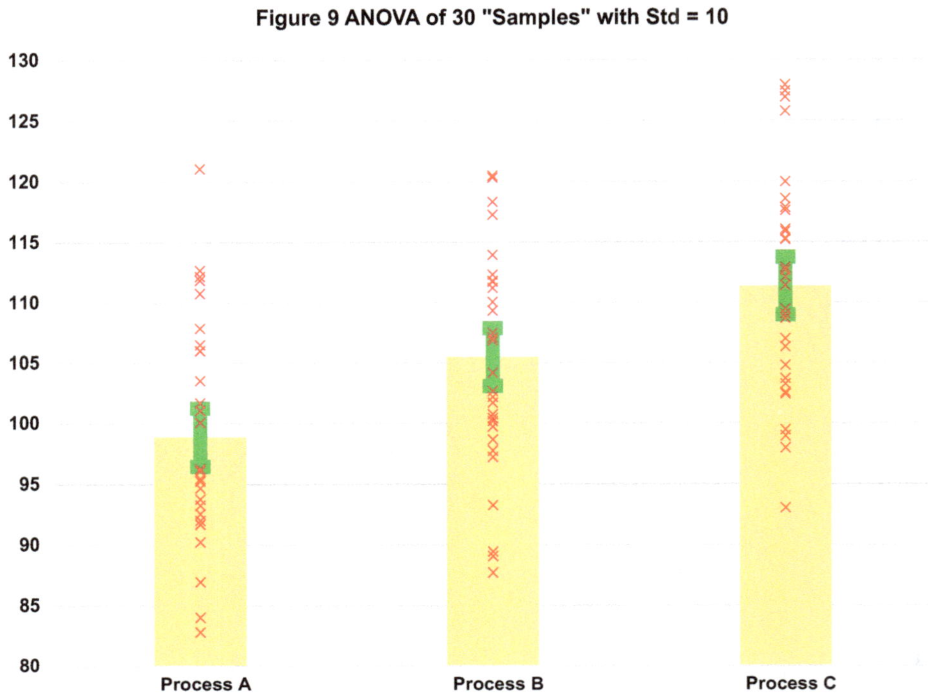

Figure 9: “CI Plot” of Example Data with Thirty “Samples”.

We often overlook this first observation. Extensive testing processes give us great confidence in AVERAGE results, but we are frequently shocked by individual experiences. Drug testing is an excellent example familiar to us all. We must always look deeper into the testing to understand how individuals respond.

The second observation confirms our intuition. So now we know that seven samples are not enough, but 30 samples are more than enough to detect differences between all three processes. We use observations two and three to estimate the number of samples needed BEFORE we run extensive testing. If we did our Intuitive Phase well, we would have data to calculate standard deviations and evaluate the “Least Significant Difference” (LSD), the smallest difference between averages that matter to us. We should choose LSD based on key business factors such as costs, benefits, liabilities, capabilities, and other development constraints and opportunities unrelated to the testing. For convenience, DesignExpert® uses LSD as the Confidence Interval in ANOVA graphs.

After we select our LSDs and estimate standard deviations, we can do a Power Analysis in a good DoE software package to calculate the number of samples we will need. If we do the Power Analysis correctly, we should have enough samples for reliable testing. For more on this technique, see an excellent text on DoE and DesignExpert®, DOE Simplified [11].

In designing our approach, we are protecting against two possible errors. In our example, we are trying to protect against:

1. accepting a new method that does not work (a "false positive"), and
2. rejecting a new method that does work (a "false negative").

Note: Keep in mind that, in our example, we have defined "work" and "not work" based on making a difference of 10 at a 95% confidence level.

In statistical parlance, these are type I and type II errors. We are said to be testing hypotheses about how these two methods compare. To use standard statistical mathematics and terminology, we must state things in strict ways that sometimes sound awkward. For example, we must state a "null hypothesis" and an "alternative hypothesis" covering all possible outcomes. The null hypothesis is framed such that "nothing changed." The alternative hypothesis covers all other possibilities (i.e., the negative of the null hypothesis). If we do this, then we can apply standardized statistical methods. If we don't follow this strict logic, we will misapply the statistical tools causing big blunders.

In our example with the microwave popcorn, we are trying to find a new method that works better. We need to approach this almost like we would in a "process of elimination." We propose a new method and then state our null hypothesis as, "Our new method is no better than the standard method," and proceed to gather enough data to prove that null hypothesis is wrong with high confidence (usually 95%). We can then conclude with high confidence that the new method is better. We accept the alternate hypothesis that "Our new method is better than the standard method."

This process may seem convoluted, but there is some method to this madness. First, statistical modeling comes from science and should be deliberately skeptical of any claim. If you cannot prove that the negation of your claim is false, we should assume your claim is false. Hence, the burden of proof is on the person making a claim. Second, the mathematics of assessing type I and type II errors is not "symmetrical." We cannot "flip" things around without changing the bias from skepticism to optimism.

Furthermore, there is no simple way to convert what we know about a type I error into a valid statement about a type II error or vice versa. Hence, if we reverse the roles of the hypotheses, we will get the wrong answers. Therefore, we must stick with the standard approach, even though sometimes confusing.

We now have some essential statistical tools to begin testing and improving technology using a deductive process. The intuitive phase of our study should arm us with considerable experience with the crucial issues and the testing protocols we will need to make our hypotheses and make careful measurements. When we say, "make careful measurements," we mean run experiments (sometimes called "trials") under controlled conditions with specific purposes in mind. We want to eliminate uncontrolled factors that will confuse our results. Hence, "making careful measurements" is usually an expensive undertaking. It often requires specialized equipment and personnel. It sometimes requires "pilot-scale" equipment that uses large amounts of raw materials, energy, human resources and may produce pounds if not tons of unusable product or waste. Since each test is costly, we want to minimize the number of tests we perform and maximize the value of each test. Hence, we will:

1. evaluate the objectives of the testing carefully and ensure the testing plan will yield relevant information,
2. plan the details of the testing carefully,
3. use the appropriate measurement tools,
4. monitor and document the testing carefully,
5. run as few tests as we think we need to meet the objectives of the testing,
6. test as many variables at the same time as we think practical,
7. test the variables at multiple, appropriate levels, and
8. evaluate the results thoroughly and objectively.

These concepts will impact how we set up, execute, and evaluate our testing plans. First, however, we must have absolute clarity on why we are doing a test, what we hope to learn, and how the results will move the program toward commercialization. We have passed the time to be less formal and even "play around" with ideas. We now must commit to progress or abandon the program. Wandering around in a fog or wasting precious resources generating useless data is inexcusable at this stage in development.

Second, we must plan tests carefully. If we lose control of the experiment, not only will the data be ruined and time wasted but we also run the risk of property damage and injury to personnel. Even laboratory-scale experiments can have serious consequences. Trips, slips, cuts, burns, and unintended exposure to poisons, pathogens, and carcinogens are common accidents in a lab setting. When we scale up experiments to pilot and semiworks, we increase costs. We also create additional hazards such as falls, explosions, fires, asphyxiation, environmental releases, and vehicle crashes. Hence, we should insist that every experiment is necessary to advance development. We should design every experiment and create a run plan that details the work. Finally, we should review those run plans to ensure that all the materials, equipment, and appropriate personnel are available and adequately trained and briefed.

> Running experiments is hazardous. We can never be entirely sure what to expect. There is no future in being sloppy or cavalier, especially when working with hazardous chemicals and conditions.

Third, we need to use the testing methods and procedures that will give us the reliability we need to meet the objectives of the tests. We must not substitute less reliable testing methods or operational procedures. We may inadvertently compromise an entire series of tests. As we have seen earlier, the reproducibility of sampling and testing procedures are crucial factors in collecting data that meet the testing objectives.

Fourth, we must keep adequate, planned, contemporary records and that test personnel be adequately trained and familiar with the details of the testing protocols. Many significant breakthroughs in science occurred when researchers documented unexpected results thoroughly. Furthermore, we often learn that many serious accidents are preventable by evaluating anomalous data and "near misses." Unfortunately, there is a tendency to rely on automated data recording in the modern laboratory and pilot plant. Computerized systems are excellent for tracking known factors with reliability and precision, not possible with manual methods. However, manual processes like clipboards and notebooks can supplement automated systems and capture the unexpected excursions that can lead to new knowledge and avoidance of unanticipated hazards.

Fifth, because testing costs are high, we should add a minimum number of redundant runs to the run plan. Our previous experience will tell us that a small number of tests will fail for a wide variety of reasons. Therefore, we must add some redundant samples and tests to account for random system failures. Furthermore, we must add additional tests to monitor system "drift" over time. Few systems are entirely stable for long periods. Randomizing trials and creating testing blocks are two methods to deal with time instability.

Sixth, when practical, we should measure and evaluate several factors at once. It is no longer necessary or even desirable to vary only one factor at a time. This approach is "old school" thinking. This strategy made sense before computers could deconvolute multiple factors and factor interactions. Today, we should design experiments to vary several factors simultaneously and use our more powerful statistical tools to optimize the value of every investigation. If we vary factors appropriately, we can simultaneously test for nonlinear interactions between factors. We can't reveal these interactions by changing one factor at a time.

Seventh, we should test factors at multiple levels (more than two) whenever practical. This approach will give insight into the mathematical relationships between input factors and results. Most experimental designs default to linear relationships between input factors and results. This assumption may be entirely erroneous and very misleading. For example, if a slight change in an input factor results in a large difference in output, we could have a significant breakthrough. Frequently, the effect may "saturate" and become very nonlinear. Multilevel factor testing will alert us to nonlinearity.

Eighth, we must evaluate experiments thoroughly and objectively. We should never arbitrarily ignore or dismiss data. Ignoring or "cherry-picking" data is the fastest and surest way to introduce bias that leads to fraud. Whenever practical, evaluations should use "disinterested" parties to avoid "expert bias" and the "explaining away" of "anomalous" observations. We should evaluate our data as soon as practical after completing a test run. The sooner we complete the data evaluation, the more the data will help guide subsequent testing. Most knowledge is cumulative. Having a thorough review of previous experiments before starting additional experiments can effectively avoid "blind alleys" and create the most relevant new hypotheses to be tested.

Rarely does our first set of trials give us completely unambiguous results. It is also rare that our first set of tests is entirely useless. Instead, we often learn something that leads us to more exacting hypotheses and statistically valid conclusions. The deductive phase is usually an iterative process that leads us smoothly to detailed, reliable knowledge. Occasionally, something novel ("good" or "bad") happens, and the experimental team must go back to the intuitive phase and learn more about the process in general.

The purpose of this deductive phase is to obtain reasonable working models of the processes under investigation with quantifiable levels of confidence. Although often not stated overtly, most statistical models are least squared, linear regression. The hope is to obtain a simple, linear equation for the relationships between factors and results. Computer programs create linear equations by minimizing the square of the difference between experimental data points and the points predicted by the linear equations. The mathematics is a little hard to describe in words, but it is straightforward and easily lends itself to analysis of variance (ANOVA). ANOVA is a powerful statistical method to compare multiple factors.

A linear model is nothing more than a linear equation that is a "best fit" for the experimental data. Recall that a linear equation is of the form:

$$y = mx + b \tag{4}$$

where y is the result, x is the factor, m is the slope of the line, and b is the intercept of the line.

The "m" term indicates the effect of a factor on a result. The bigger the "m," the more a change in "x" creates a shift in "y." The "b" term is an "offset." It is the theoretical value of y when $x = 0$. Statistical models include another term, "e," a random error term. The "e" expresses the difference between the data points predicted by the linear model and the actual data points. If the sum of the squares of "e" is small, then the fit is very good. When there is more than one factor, the equation takes the form (here we include multiple error terms):

$$y = m_1x_1 + m_2x_2 + \cdots + m_nx_n + b + e_1 + e_2 + \cdots + e_n \tag{5}$$

The linear model immediately infers statistically valid relationships between factors and results. If the "x" term is an actual cause (not just random correlation), we might even discover the physical cause of the result. If the "e" terms happen to be "normally distributed" (a mathematical definition), we can apply ANOVA calculations and get additional statistical information. We can get:

1. the R^2 values, which indicate the goodness of fit for the model;
2. the adjusted and predicted R^2 values which indicate the robustness of the model; and
3. the p-factor for the F-value means the statistical significance of various factors (i.e., which ones are significant for the model).

These interpretations can be powerful, but they can also be deluding. The model assumes linearity, and ANOVA works only when the error terms are "normally distributed." This assumption is far from correct in many cases, and the linear model is not very useful. Often the relationships between factors and results follow nonlinear physical laws. When this is the case, sometimes the data can be transformed (often with a power or logarithm calculation) and made to fit the linear model. The interpretation of transformed data can become very complex.

The simple linear model we have shown also assumes that the factors do not interact. Unfortunately, this assumption is often not the case – especially with physical processes. When factors do interact, we can expand the linear model to include interacting variables, but the physical interpretation of the model becomes much less meaningful. However, linear models using the products or ratios of factors can be helpful over the range of interest, even if the relationships are not strictly linear.

From the above, we can see that we should be careful when interpreting the results of a linear model in some physical way. If we are very familiar with the system, the coefficients and the intercept might have physical meaning. It might even indicate what physical laws are involved. On the other hand, we should use the model as empirical curve fitting if we know very little about the physical laws affecting the system under test. When that is the case, we may have a useful model, but it tells us very little about what is happening physically or chemically. It would be foolish to extrapolate such data into untested areas. This observation can be generalized by:

> Because we rarely know with much confidence all the physical laws governing a system under test, it is foolhardy to extrapolate data outside our experimental conditions.

Extrapolating a model outside of experimental conditions is a frequent source of those "major blunders."

Guarding Against Major Blunders

There are many ways to make significant errors in our testing and analysis. They generally fall into one of five categories:

1. Extrapolation of data (mentioned above)
2. Sampling errors of all kinds
3. General paucity of data (drive for "efficient testing")
4. Logic errors from interacting probabilities (e.g., "Bayes rule")
5. The black swan

We have talked about item 1, but we should repeat it here. It is never good to use a model or apply results outside of the test conditions. Furthermore, when extrapolation is necessary (e.g., scaling up a pilot plant), it is effective only if the significant underlying causal relationships are well known. An example might be trying to scale up a thoroughly understood reaction. If, for example, we have solid data that reaction temperature is the critical element of a successful reaction, and we know the heat generation and dissipation mechanisms of our larger reactor, then we have a reasonable chance at predicting the performance of a larger reactor. On the other hand, if we do not have that kind of detailed information, it is perilous to assume that the process will scale up well to a larger reactor size.

Sampling errors are the kiss of death to many Technology Development programs. Sadly, most programs only require "representative" samples without specifying what that means. Occasionally, sampling is trivial. It is only a matter of taking a portion of a well-mixed stream. The occasional success we experience with simple random sampling only lures us into complacency. Far more often, representative sampling is difficult and often nearly impossible. Often specialized sampling equipment designed for the specific application is necessary to obtain barely adequate samples.

Sampling is a complex issue that we cannot adequately cover here. "Correct sampling," however, must be considered at the start of any program. The quality of our sampling approach also needs to be validated before final reporting. "Correct sampling" is first and foremost unbiased sampling. That means that every possible increment of a lot population should have an equal probability of being sampled from the main population and any subsets selected for analysis. Sample bias is tough to detect after the fact. Hence, we must design correct, unbiased sampling into the process. Particular attention should be given to the expected heterogeneity of the population and compared to the actual results obtained from the sampling system. If unexpected sample scatter is observed, "incorrect sampling" should be suspected and further investigated.

For the most part, all testing programs suffer from a lack of data. We always want to get the most information from the fewest experiments possible. Determining what is "just enough" is not a simple task, and much of DoE focuses on solving this puzzle. We can generate a linear model for n factors with $n + 1$ data points. Unfortunately, this gives us no information about the experimental error. Hence, we have no good way to judge the model's reliability. We must always add mathematically "redundant" data points to give us information about model reliability.

As we end the deductive phase, we need to review the logic we used in hypothesis testing. If we have followed the sound, conservative research principles, we will have set our null hypotheses to negate what we would like to prove. Our testing biases will be against accepting claims unless the data overwhelmingly favor that claim. We will have designed our testing protocols to detect all relevant input factors that significantly impact the outputs of interest. Furthermore, we will have determined the relationships between key input factors and relevant results within reasonable error. Where practical, we will have developed valid models to evaluate and optimize the value of the technologies under study.

As we approach the conclusion of our deductive phase, we need to begin to pull together all our learning. We certainly need to check our logic, methods, sampling systems, and objectivity. We are about to ask our organization or others to invest large amounts of resources in the next commercialization phase. Here, we want to do at least one set of redundant tests to verify our findings – especially any models created. This step is often called a "validation" step. We have generally set a 95% confidence limit for much of our decision-making in our testing methodologies. That leaves us a 5% chance that we are just dead wrong by the luck of the draw.

If we start from scratch and do a completely independent verification of our initial conclusions, we will reduce our chances of a complete blunder to 0.25% or less. If we have developed a model, we should use it to predict the outcome of a test run at several different operational points within and at extremities of the operation conditions. When asking for significant investments, we need this assurance that we are on the right track. Too often, executives fail to take this single, critical, reasonable step before committing the organization to a course of action that can result in its demise.

In our final review of the development process, we need to look for logical errors that may have crept into our process. Of course, we need to review our assumptions and hypotheses, but we also need to look for any errors related to conditional probabilities. A conditional probability occurs when one of two related events has occurred, and we want to predict the probability that the other related events will occur. This situation is prevalent in business decisions where we have multiple related events naturally occurring in succession. If the events are entirely independent, we generally get our logic right. However, our intuition often leads us astray when the events are related (as in a conditional probability). We will not go into great detail here, but we will illustrate with an example.

Suppose that extensive studies have shown that about 1% of the coal cars we receive have unacceptable sulfur content. Also, suppose we have a simple test for high sulfur, but it is not perfect. And finally, assume that we have good studies that show:

1. 80% of coal cars that have high sulfur test positive for it by this simple method and
2. 9.6% of coal cars that do not have high sulfur also test positive for high sulfur (i.e., "false positive").

Now suppose that we have a coal car that tests positive by this simple method for sulfur. Should we immediately shuttle the car over to the high sulfur treatment facility? Part of the answer lies in answering the question, "What are the chances that the coal car actually has high sulfur?" Think about that for a moment before going on. What would your guess be?

The correct answer is that there is only about a 7.8% chance that the coal car actually has high sulfur. Unfortunately, most people (including many scientists and engineers) get it wrong and guess a much higher percentage. We can calculate the correct answer with Bayes' theorem for conditional probability, but it is complex and not very intuitive.

There are simple ways to approach these problems without digging out the statistics book. If we carefully catalog all the various possibilities, we have a better chance of getting the correct answer. Not only does it help us organize our data, but it also helps us determine if we have enough information to solve the problem.

> There are many logic problems in Technology Development. We must keep our logic straight and not be carried away by our "intuition" – especially when we have our ego invested in the outcome of the decision.

Example 15: The *Monte Hall* "Paradox"

Monte Hall was an American TV Game Show host. At the end of the program, the contestant was allowed to choose one of three doors to win their final prize. The top prize was often a new luxury car. We modify this scenario a bit to illustrate a logical dilemma. Suppose that after the contestant had chosen one of the three doors. Let's say they chose door #2. Now suppose they were allowed to choose a trusted person to go backstage and look behind doors #1 and #3 and tell the contestant which one of the remaining doors did NOT have the luxury car. Of course, it would be possible that neither door had the luxury car, but we only allow the trusted person to say that one of them did not have the car. Suppose the trusted friend said door #1 did not have the luxury car. The question is, with this new information, should the contestant change his choice to door #3?

Our intuition here is poor. Many think it doesn't matter, and the contestant might as well stick with their first choice – door #1. Others feel that the contestant should change their choice to door #3 because the chance for success is better now. Many believe that the chances of being correct are up to 50/50. After all, the contestant is currently choosing between two rather than three doors. Few get the answer entirely right.

The correct answer is that the contestant should always change to door #3 (or whichever door is not named by the trusted friend). The contestant would double their chances of winning the luxury car from 1/3 to 2/3!

The reader is encouraged to verify this by experiment. Use three playing cards, one representing the luxury car (we'll call it the "prize" card) and the others the less valuable prizes. Shuffle the cards and place them face down on the table. Pick one as your choice, but don't look at it yet.

Ask a trusted friend to look at the other two cards you didn't initially pick. Have them tell you one of those cards that is not the "prize" card. Run the game for at least 20 cycles sticking with your first choice. Your success rate will converge to 1/3. Repeat the experiment, this time always changing your selection to the card you did not choose, AND that your friend did not say was NOT the "prize" card. You will see that your success rate will approach 2/3.

Why is this so? It only makes sense if we carefully think through (i.e., enumerate) the possibilities. If you make no change with the new information from your friend, you have a 1/3 chance to select the "prize" card. But there is a 2/3 chance that the "prize" card is among the two cards you did not choose. When your trusted friend tells you which of the two cards is NOT the "prize" card, then 2 out of 3 times, the "prize" card will be the other card that your friend did not mention.

Why is it hard to "see" this? We have grave difficulty in getting our heads around the consequences of sequential choices. Try to explain the "experiment" to someone else after doing it. You will see that it is not easy to explain what to do. Our brains don't function like that.

For all our care and planning, everything may yet fail. Because we don't know what we don't know, we cannot protect against every contingency. In the landmark book, *Fooled by Randomness* [27], Dr. Nassim Taleb resurrected the idea of the black swan. The black swan is an important event so rare that it is "never" observed. According to Taleb, a black swan event has three characteristics:

- It is highly improbable.
- It is highly impactful.
- We imagine irrelevant factors "explaining" the event that makes it seem more probable than it is.

There is nothing we can do with black swans. Because they are so rare that we cannot frequently observe them, we don't even have the opportunity to learn about the real black swans. We can only recognize that they are possible and avoid the trap of inferring too much from rare events and limited data sets.

Execute Your Plan

So far, we have been talking about the technical elements of technology evaluation and development. We need to know these. We need to thoroughly understand how our strategies relate our technology to markets, develop our hypotheses in an intuitive phase, apply the science of experimental design to our hypotheses, and verify our results and logic. This process is quite flexible and applicable to a wide variety of problem-solving situations. It is not a haphazard "knocking about" hoping for divine intervention. It can and **must** be mapped out and tracked. Technology Development should be planned and executed like any other purposeful human endeavor. Anyone who thinks that Technology Development is unmanageable completely misunderstands the nature of knowledge and the processes of accumulating knowledge. The error bars around the timing of the individual steps are certainly larger than those surrounding the construction of the 35,001th McDonalds. So what? That only means we should plan, evaluate, and implement the process with more excellent skill, discipline, and art.

Of all the many responsibilities of the executive, their most significant contribution will be in the **execution** of a Technology Development program. Of course, the executive should insist on written plans that contain milestones, critical paths, estimated resources, PERT projections, formal reviews, documentation procedures, and a host of other important planning and organizing aspects. But the most significant contribution of the executive will be controlling the process so that the Technologists faithfully execute the plans and honestly record and report results. Controlling Technologists is not a trivial task. There will be "failures" in the sense that things will never go as planned – at least not precisely. There may be some embarrassingly awful news. Adjustments will be needed, and there will be the temptation to dodge the news rather than deal with it honestly and effectively.

The executive should take the following warning:

> Most Technologists will chafe under a well-managed Technology Development Program.

Some of this has to do with personality. Many Technologists are creative and want to follow their curiosities without the constraints of organizational objectives. Other Technologists will chafe under the structure out of fear. Some (especially the more senior) are **not** the experts they claim to be. Hence, frequently their reservations are more about ego-preservation than creativity. Some Technologists are just plain lazy. Here the executive can often reassure and encourage. The executive **must** have excellent interpersonal skills **and** an understanding of the personalities of the Technologists. We will return to this theme a little later in this book.

Evaluation and Reporting

Before leaving the deductive phase, we must mention the need to evaluate and report our findings. We have talked a bit about doing the final verification of our results. We have also talked about the need to review our logic. However, we need to go at least one step further. We must evaluate our findings given our hypotheses and our overall strategies. This evaluation must include the impact of our results on our business goals. In other words, we need to take yields, production increases, cost changes, and so on, and propagate them through our many financial and business assumptions to see the impact on the markets we have established in our strategies. In other words, we need to expressly state what we think we have learned as it pertains to our business goals.

> Another way to put this is to ask the questions, "So what? . . . What next?"

Let me illustrate with an example. Let us suppose that we want to save money on a chemical process. By substituting a less expensive raw material, we might cut our overall costs. Before we go very far into running any lab or pilot tests, we need to define what we mean by "saving money." We probably cannot test "saving money" directly in our lab or our pilot plant, and no one in their right mind would turn an operating plant over to a bunch of geeks without guidance on what to do. Hence, we would have to gather pertinent data from lab- or pilot-scale work and predict the impact on total costs at a plant level.

To make any real progress, we would need to put all the testing results into a mathematical formula that expresses the relationship between potential raw material cost savings and other effects found in our testing. If, for example, we found that changing raw material reduced yield, we would want to know by how much. Likewise, if we found that the change increased the probability of having severe plant failures, we would need to have some way to factor that into our assessment. Of course, we would want to know how all these factors worked together on average, but we would also like to know how they impacted the variability of results. Variability of results means risks, and that translates into additional costs. If changing to a new raw material ruined the reliability of our plant, we need to know that.

If we can write down a mathematical equation to calculate our final result, we have a "model" for our process. Some might complain that we really cannot do that until after our experimentation. This thinking is flawed. How factors impact results is always a necessary hypothesis. Failure to state overtly or make provisional hypotheses leaves the experimental design with no reasonable basis. Why would we ever do experiments if we were clueless on how the results might impact our business? We must commit to some guesses and start testing them or get nowhere. We can always start with simple hypotheses and linear models and then proceed to more complexity as we learn more.

Once we have an overall working model, we can design a comprehensive strategy for testing hypotheses. Furthermore, we can assess the variability and uncertainty in results caused by the variability of the raw material costs, yield, quality, downtime, and other factors. Finally, we can assess the uncertainty created in our final results from the error in our measurements and methods.

Evaluating how the uncertainty and variability of individual factors impact a final derived result is known as "propagation of error." We can propagate the error for most linear models and many nonlinear models if we have enough data. We will hold off on the mathematical details and understand how this technique works and why it is essential. The illustrations used here are kept simple.

Let us suppose that we have found a way to predict the average cost of a new raw material (C_M) and the average manufacturing and distribution costs (C_P) that would arise from using the new raw material. Then, our predicted total average costs (C_T) would be

$$C_T = C_M + C_P \tag{6}$$

We would, of course, insist that this new C_T be, on average, lower than our current C_T. It is nice to know the average or expected value of C_T, but we also need to see the uncertainty around this new C_T. We will go more into the math later, but it turns out that in this type of equation, the variance of C_T is equal to the sum of the variances of C_M and C_P. That is:

$$\mathrm{var}(C_T) = \mathrm{var}(C_M) + \mathrm{var}(C_P) \tag{7}$$

We now have a way to assess the uncertainty of this new cost structure if we can measure or predict the uncertainty around C_M and C_P. We might see, for example, a slight improvement over our existing cost structure, but we might also see a lot of uncertainty around that. As a result, we might hesitate to implement such a change, especially if we were confident that our current cost structure is very predictable and profitable.

Equation (7) rolls up a lot of experimental data into just a few simple variables. In practice, the experimenter is working at a much less grandiose scale. A more realistic case for an experimenter might be assessing the uncertainty of chemical yield. The experimenter might run many trials to measure the variability of the yield. There are many occasions where "propagation of error" calculations can be very useful before and after experimentation.

We might want to do an a priori calculation (probably better called an estimation) to help us evaluate which factors are likely to be the most important to the precision of *future* measurements. Some errors are multiplied dramatically by how they impact final results. Propagation of error calculation alerts us to the sensitivity of the measurements. This kind of calculation might reveal that, under current conditions, we cannot possibly meet the precision requirements of our project. We might need to redesign some of our analytical equipment before wasting money on hopeless trials.

We might also use this type of calculation a posteriori (i.e., after testing). We may do this to validate our assumptions. We may also wish to combine the impact of tested factors with those we did not test. For example, we often want to estimate performance when scaling industrial processes from lab or pilot plant data. Here we might have to use "expert" estimates (often called "SWAGS") or secondary data in our business modeling for factors that we could not measure directly. Propagation of error analysis can bring together disparate information giving objectivity to potentially complex analysis.

Let us look at an example more typical of what we would see at a pilot plant. Suppose we want to measure the yield in our pilot plant. Yield is often a result that we derive from several independent measurements such as:

- The mass of mixed product produced (P)
- Compositional analysis of the mixed product (wt% A)
- A mass of mixed raw materials (R)
- Compositional analysis of the mixed raw materials (wt% B)

Let us also suppose that the key yield indicator is the mass of A in the product per mass of B fed in the raw material. Our yield formula (as a fraction) would be

$$\text{Yield} = \frac{\text{mass}\,A}{\text{mass}\,B} \tag{8}$$

But we do not weigh A or B directly. We calculate it from

$$\text{Yield} = \frac{\text{mass}\,P \times \text{wt\%}\,A}{\text{mass}\,R \times \text{wt\%}\,B} \tag{9}$$

Now suppose we want to estimate the error in the yield before doing any experiments. It turns out that this type of error propagation is also reasonably straightforward. When formulas are simple ratios, the square of the coefficients of variation is additive. The coefficient of variation (CV) is the standard deviation divided by the mean. Hence

$$(\mathrm{CV}_{\mathrm{Yield}})^2 = (\mathrm{CV}_{\mathrm{mass}\ P})^2 + (\mathrm{CV}_{\mathrm{wt\%}\,A})^2 + (\mathrm{CV}_{\mathrm{mass}\,R})^2 + (\mathrm{CV}_{\mathrm{wt\%}\,B})^2 \quad (10)$$

Formulas like eq. (10) can be handy for helping design and interpret experiments. If we know, for example, that we will have a difficult time measuring the wt% of B in the raw material, we know that the variability/uncertainty around yield will be high before even doing the testing. Thus, we might want to get a better way to measure B before we do a lot of testing. On the other hand, if we have good data for mass measurements from our vendors of weigh belts, load cells, and so on, we might conclude that it would be a waste of time to check and recheck the mass measurement systems. Instead, we could verify that they are working as designed and focus our attention on the more difficult measurements (such as measuring wt%).

As we have said earlier, it is almost always the case that we have only part of the experimental data we need to fully "prove" a complex business proposition. For example, we may have empirical data for yield. We might even have experimental data on failure rates, product quality, and so on. Nevertheless, we will almost always have to make guesses or extrapolations rather than measure the bottom-line numbers like profitability or total costs.

If we can state the result of interest in a mathematical formula, we can usually develop a general formula for error propagation. This approach may help us predict the variability of our results of interest even when we are short on experimental data. The general approach to error propagation is described mathematically in the following paragraphs and equations. The following discussion is from John Mandel's book, *The Statistical Analysis of Experimental Data* [24].

Let $x, y, z, \ldots$ represent random variables whose true values are $X, Y, Z, \ldots$. Let u represent a derived quantity whose true value is given by

$$U = f(X,\ Y,\ Z, \ldots) \quad (11)$$

Let $\varepsilon_1, \varepsilon_2, \varepsilon_3, \ldots$ represent the statistically independent, small errors in $x, y, z, \ldots$, respectively. Then the error induced in u, which is denoted as ξ, as a result of the errors $\varepsilon_1, \varepsilon_2, \varepsilon_3, \ldots$, has a variance estimated by

$$V(\xi) = \left(\frac{\partial f}{\partial X}\right)^2 V(\varepsilon_1) + \left(\frac{\partial f}{\partial Y}\right)^2 V(\varepsilon_2) + \left(\frac{\partial f}{\partial Z}\right)^2 V(\varepsilon_3) + \cdots \quad (12)$$

Equation (12) looks daunting, but when the formula for the result of interest is a mix of sums and products, the evaluation usually is not too complicated. It requires successively taking the partial derivative of the model equation relative to each variable of interest while treating the other variables as constants. We need an estimate for the uncertainty for each factor at or near the operating point of the model. Fortunately, there are now excellent multivariant data analysis software packages to assist in these calculations when they become complex. For more information, see the excellent book, *Multivariate Data Analysis, 6th ed.*, by Esbensen and Swarbrick [25].

Although the analysis can become complex, evaluating the total error of a process is an indispensable technique. It is essential to have a reasonable and unbiased estimate of the uncertainty of any result that will impact a business proposition. By its very nature, business turns on the issue of uncertainty. Companies can often command a premium for handling uncertainty effectively. High stakes can be very rewarding. Nevertheless, those who find themselves with the wrong mix of uncertainty and reward will inevitably fail.

We must document the data/trial evaluations described above in writing. All too often, the conclusions, ideas, recommendations, anomalous observations, additional hypotheses, caveats, and other vital information get lost to the organizational knowledge base simply because no one wrote these things down. The executive cannot let this happen. There is hardly anything more valuable or volatile than the organizational knowledge base. Therefore, the responsibility for recording the experiences and results of the trial should be made very clear at the outset.

Formal reports are almost universally hated. No one seems to want to write them, hardly anyone will voluntarily read them, and managers seldom wish to set aside resources to produce or archive them. And yet, there is hardly anything easier to do or of greater value than writing down, in succinct terms, what the heck happened.

Formal reports should be required. The executive must identify the person(s) responsible for writing the formal report and those who read and approve the final report as part of the study or trial. The executive should never leave these mundane, thankless tasks to “anybody” or “somebody.” The persons assigned should be knowledgeable, engaged, and responsible for outcomes. The work must be reviewed and approved by persons not doing the work. Reviewers easily find omissions and contradictions that persons who performed the work overlook.

A formal report on a well-designed “study” or a “trial” should be straightforward. If not, then the study or trial probably had severe design flaws. Often, a useful report contains just a few pages. All such reports should follow an outline something like:

1. Abstract/introduction – Generally, what was done and why?
2. Experimental – Specifically, what was done? Give enough detail so that others could repeat the testing or trial.
3. Results – What were the results of measurements?
4. Discussion – What results were expected? What results were unexpected? What, if anything, was or might have been “out of control?” What are some potential reasons for factors being “out of control?”
5. Conclusions – A summary of what was learned – if anything.
6. Next steps – Repeat. Revise. Move on to something else.

All reports should be:
- written and stored in a searchable, electronic format,
- written in a consistent format/organization,
- keyword indexed,
- security protected to an appropriate level, and
- archived for reliable retrieval.

Additional guidance can be found in ISO 9001 [12] and ISO 17025 [13].

Not all researchers will agree with the consensus on some trial results. As a result, legitimate disagreements on future testing priorities can arise – especially when faced with unexpected results. The executive should encourage recording "minority" opinions since they could be the kernel of discoveries.

Improve Your Experiments or Wrap Up Your Conclusions

After reporting the results of the trials, there is still a significant decision to be made. The question is, "Are we done?" The answer is never simple. It depends on many business considerations. It could be that we have proven the key hypotheses with sufficient confidence, and the results are "good enough" to warrant moving to the next step of commercialization. On the other hand, our results might be "encouraging," but we lack sufficient confidence to move on. We would then re-evaluate our trial methods and redo some or all of our testing to see if we can verify our "encouraging" results.

We might find that the confidence is reasonable, but the benefits do not look "good enough" to proceed with the next commercialization step. We would want to evaluate the gap between performance and potential improvements to that performance. We might have some specific ideas to try. If that were the case, we would want to design a new testing program with a new set of hypotheses that could "improve results."

There is always the chance that we don't know what to do next. We might see no obvious way to close the gap between our results and what we need for business success. This situation occurs far more often than researchers would like to admit. They are not exactly the most optimistic folks on the planet, but they are certainly among the most tenacious and unlikely to give up. The executive must tread the squishy ground between plausible and practical. They can take input from the Technologists but cannot abdicate the fiduciary responsibility for making the right call.

One final comment is needed. The reasoning behind the decision to move to the next stage, re-engage in testing or abandon the program should be thoroughly documented. Business conditions change. New evidence or processes emerge. Clear documentation of the reasoning behind a decision will prepare the organization to react quickly to changes in the business landscape. The organization could pick up the program where it left off and gain a tremendous competitive advantage on those who might have to start their development at an earlier stage. In this respect, technologies are continuously under development.

Valuation Methodology (Present Value/IRR)

As we noted earlier, there are many different schemes for valuing an investment. Much depends on the investor's investment strategy and perceptions of risk. For early-stage, technology-driven companies, enterprise valuation can be challenging. Simple "rules of thumb" can be inviting, but they are nearly worthless. An example might be the "Berkus method" presented in *Winning Angels* [14] by Davis Amis. It can be summarized by

If exists	Add to company value up to
Sound idea (basic value)	$1/2 million
Prototype (reducing technology risk)	$1/2 million
Quality management team (reducing execution risk)	$1/2 million
Strategic relationships (reducing market risk)	$1/2 million
Product rollout of sales (reducing production risk)	$1/2 million

An immediate problem for an early-stage, technology-driven company is determining if the idea is sound. Hence, the "Berkus method" is not helpful since we can't get past the first step. The remaining steps are somewhat beneficial in thinking about value, but they are irrelevant without sound ideas. So, what is a "sound idea," and how would we know one if we had one?

The first step is technical feasibility. Most, but not all, ideas floating around are technically feasible. That is, they would work. Some ideas – even some that get funding for millions of dollars – never had a chance. They are, of course, completely bogus claims and horrific blunders in testing. A significant number of these crack-pot ideas even have associated patents.

> Just because something is patented, it doesn't necessarily work – at all.

Often, it is not that the idea doesn't work at all; the problem turns out to be that there is no way to make money with it. Commercial feasibility is usually the most significant issue. Hence, the valuation of every early technology eventually boils down to evaluating the feasibility of making money. Rather than dodge this fundamental problem and substitute some bogus "rule of thumb," we stick with a more direct approach. The analysis results are the departure point for negotiations between the offerer and the potential investor.

We use discounted cash flow projections as the basis for valuation discussions. We include investments made as negative cash flow (including planned subsequent investments) and estimated profits as positive cash. Since this is all based on estimates, we do not consider depreciation or tax consequences unless they are significant, planned, relatively certain investment events (e.g., statutory or negotiated tax relief). We frequently run leveraged and unleveraged models for comparison.

We generally calculate the IRR as our first cut of the modeling. We do this for two reasons: (1) makes no presumption of investor risk tolerance, and (2) many investors have a minimum IRR in mind (which they often "adjust" based on perceived risk and other subjective factors). We generally prefer to talk about the IRR of the project to keep the focus on the potential earning power of the technology compared to the cost of creating that earning potential.

In truth, we are reluctant to provide a numerical value for the company – especially if they are in the early stages of Technology Development. It is frankly a very uncertain number, but it is one that everyone seems to want and will calculate a value on the back of a napkin if given a shred of data. These are the MBA types who can't bring themselves to admit that their "investment" in an early-stage company is more like buying a lottery ticket than an asset. The key is understanding the potential payout, the probabilities for success, and seeing if various scenarios match your appetite for "playing the game."

Though we are reluctant to do so, we frequently set values by a specific process. Therefore, as the first cut, we will use the data developed for the IRR and use the following rates with the result for valuation derived from the net present value (NPV) at a set "discount rate."

Technology phase	Discount
Idea	100%
Literature supported idea	50–100%
Bench-scale demo	35–50%
Pilot-scale demo	25–35%
Customer sales	15–25%

The result is similar to the "venture capital model" presented by Montani et al. in "Startup Company Valuation: The State of Art and Future Trends" [15]. Note that an idea standing alone by itself has no value. It is just a dream without some justification to support it.

This "NPV model" forces the offerer to produce a detailed plan that itemizes:
- timing, costs, and milestones of the Technology Development process;
- timing, costs, and milestones of "monetizing" the technology (whether building a plant to produce a product, licensing the technology to someone else, merging with a strategic partner, etc.);
- timing, markets, revenues, and customers to serve; and
- details of the "ask" (how much? When?) to the potential investor.

In addition, the "NPV model" serves the multiple purposes of:
- forcing the offerer to deal with the details of making money on their idea;
- revealing the "cool" ideas that have no commercial value;
- showing many of the assumptions, steps, and risks surrounding the process;
- clarifying what the offerer wants from the potential investor and how much the offerer is willing to "share" with the potential investor;
- estimating the actual earning (IRR) potential of the process;
- initiating detailed risk/benefit discussions between the offerer and potential investors;
- revealing the true interests and risk tolerance of potential investors;
- providing the raw data needed for investors to run their modeling; and
- providing a detailed plan that can serve as a benchmark for subsequent go/no-go decisions.

The "NPV model" also focuses on the primary financial market transaction rather than secondary market transactions. The focus is on using financial assets to create and develop intellectual property that has wealth-creating potential. Most other approaches attempt to "guesstimate" what values investors might get in the secondary financial markets and somehow factor that into the valuation. Occasionally, one of these "rules of thumb" will work, spawning a new book that ignores the hundreds of failed predictions. Unfortunately, using bogues "rules of thumb" is an inefficient way to use capital that stifles economic growth. Furthermore, it encourages investors to buy and sell based on "sizzle" rather than fund technologies that work.

The "NPV model" favors and encourages "full disclosure" throughout the Technology Development process. This approach lends itself to matching offerer and investor expectations and risk tolerance by stating risks and rewards upfront. It should lead to the early abandonment of poor projects with serious, irreparable flaws. It should also lead to some measure of adjustment and recovery of projects with repairable flaws. And finally, the "NPV model" encourages "partnering" between the offerer and investor to achieve the mutual goal of developing a new, useful technology that has sustainable wealth-producing potential.

Example 16: Liar's Poker
I will never forget one of my earliest consulting gigs. I was helping a small laboratory put together their offering to potential investors. The owner had "puffed" the sales and profit potential claims far beyond reasonableness. Like all clever lies, it was "supported" by detailed, subsidiary claims that were complete fabrications that would be difficult to verify. When I pointed this out, I got the pat-on-the-head response, "That's what investors expect – lies. Never tell a potential investor the truth. They will discount everything you say." After weeks of painful negotiations with the "investor," we learned they had no money to invest. They had been trying to wrangle a "deal" to get control of the laboratory by "trading" worthless paper for what they claimed was something of value. A few weeks after that, the laboratory closed. Other "lies" had come home to roost.

The "enterprise valuation" generated by the discounted cash flow projection is another "living document." Over time the estimates and risks will become more evident. The potential returns may not change much, but investor assessment of risk should change and may change quite dramatically. By updating this valuation, honing the offering, and keeping in contact with potential investors, investors who may have declined earlier may decide that the "deal" isn't that bad after all.

Evidence-Based Decision-Making and Data Science

We cannot leave the topic of the science of Technology Development without discussing the strengths and weaknesses of "data science" and its application to Technology Development. We deliberately place this discussion at the end of the chapter on tools and just before the chapter on art. It will become clear that we think that "data science" has a lot of art to be effective.

First, we will need some practical, working definitions. We will try to cut through all the hype and obfuscation and get to the kernel of the issues. Among the hottest buzzwords in the industry today are:
- data science
- big data
- data mining
- statistics/data analytics
- modeling
- artificial intelligence (AI)/machine learning

But what do they mean? Are they new, space-age discoveries that only a PhD in statistics or computer science can understand? Not really. Like so many things, a maze of jargon hides the truth. The concepts are simple, and the personal computer handles the calculations. First, however, the user needs to understand the approach's limitations.

The personal computer, the Internet, and the rise of many cheap methods for capturing data and using large databases have changed the landscape. It is now possible to capture incredibly detailed measurements about most processes almost instantaneously. As a result, we now obsess over collecting data – mainly because we can. Data science has recently come on the scene to answer the question, "Can we do anything useful with all these data?"

Data science has been around since the Egyptians first started tracking the times and duration of the Nile flood and a bunch of other miscellaneous data. They first gathered information about the flood, crop practices, crop yields, and the movements of the stars and planets. One of the first data scientists noticed that when the Star of Isis (Sirius) appeared to rise just before the Sun, the Nile floods would start soon. Farmers planned their activities accordingly, and the Egyptians flourished even though they lived in the center of a vast, dry wasteland. Therefore, history indicates that data science was around by 3285 BCE.

Now, it may be a bit of a stretch to call the Egyptians "data scientists." They added lots of religious and cultic notions around their "science." Nevertheless, occult and superstition clouded data science until the experimental, scientific revolution that began in the fifteenth century (or so). Galileo became the first experimenter that concerned himself with data accuracy and precision in an objective way around 1600 AD. After that, hard sciences of chemistry, physics, and astronomy pushed forward scientific data analysis. Later, botany, genetics, and agronomy moved data analysis into areas where understanding random, experimental "error" became critical.

Over time, it became clear that statistical analysis of experimental data was crucial for understanding the usefulness of the data. Initially, scientists attributed observed variability in their data to unavoidable errors or limitations in measurement technologies. However, by the early twentieth century, many scientists concluded that all systems displayed probabilistic behavior, especially submicroscopic (e.g., photons, electrons, etc.) and complex biological systems (e.g., mutation/variations, psychological trends, preferences, etc.).

We will return to these probabilistic issues later, but for now, be aware that the mathematics of data management – especially statistics – was a well-developed field long before we were talking about big data. Moreover, data science has borrowed much of its analytical approach from the physical, biological, and social sciences.

Data science is essentially the:
- collection of data,
- the categorizing of these data,
- the creation of predictive models from the categorized data, and
- the application of the predictive models for useful purposes.

The various buzz words tend to focus on some aspect of this process and emphasize specific tools.

Two closely related terms, big data and data mining, began to emerge around 2000 and by 2010 had become **the** hot items in IT. Data storage capabilities grew dramatically during the 1990s. By the early 2000s, developments in parallel processing and large, volatile memory chips had markedly increased the speed of data searching and handling. For example, it became practical to investigate gigabytes of data simultaneously in fast, volatile memory rather than combing multiple magnetic or optical discs. In addition, it became common to perform rapid statistical analysis on gigabytes of data hunting for potential relationships between data elements, including many simultaneous variables in linear and nonlinear models.

By around 2010, the collection, storage, retrieval, and investigation of gigabytes and terabytes of data became known as big data. The methods for collecting, storing, searching, and retrieving relevant data from existing data became known as data mining. When we talk about data mining, we focus on the challenges of retrieving useful data from various independent sources. Frequently, these sources are "legacy" databases created by old hardware and software. Sometimes these databases have little structure and are of questionable quality. The data mining expert deals with all these potential issues when sorting through giga- and terabytes of data.

Two other buzz words used interchangeably are statistics and data analytics. They are not the same. Statistics is a body of scientific, mathematical principles and theorems based on various types of data randomness. It first developed from studying games of chance (dice, cards, etc.) but became closely associated with the experimental sciences (especially agronomy, botany, and chemistry).

Data analytics uses statistics extensively. Data and trends are studied to see if "statistically valid relationships exist." The question is, "Are there persistent relationships that cannot be explained simply by random chance?" The hope is that we can find persistent relationships between data elements. Ideally, these relationships are so "persistent" that they are "laws" that "cause" the observed results. We start by looking at "correlations" between data elements and then test them to see how "important" and persistent they are. In the hard sciences, we can often run enough controlled experiments that we can start talking about "causality." Finding "causal relationships" is much less common in business and the social sciences where we often are working "in the wild" or with "secondary" data – that is, data we have "inherited" and not generated under conditions that we controlled.

Data analytics is more than just using statistical analysis to find correlations. It includes evaluating the completeness, relevance, and reliability of the data. This analysis has been one of the most significant challenges for data analysts. Historical data is never complete or without obvious and suspected errors. Data analysts must decide what to do with such data without biasing the data set. There is a constant tension between determining which data are "good enough" and which data to reject as incomplete, inaccurate, or irrelevant. The data users should also be aware of the "judgments" that data analysts must make. Those "judgments" could make the analysis and conclusions biased or irrelevant.

> Anyone who thinks that Data Science is purely objective does not understand it. The Data Scientist must "prejudge" many aspects. We can only hope that it does not lead to intentional or unintentional prejudice.

The goal of data science activities is to "learn" in the sense of building and validating models that allow the measurement of readily observable factors to predict actual results. This notion is worth spending time pondering. When we "learn," we create "models" in our heads that organize distinct observations and relate them to each other. We hope that A causes B with enough reliability to get ready for B when we see A.

An example might be using data we have collected about buyer preferences. Assume we survey customer preferences about how long our product lasted. Suppose that our competition extended their warranty from 90 days to a year. We might wonder if we need to extend our warranty program. We would be scrutinizing our data to see if we could create a statistically valid model of how our customers will react to our competitor's actions. If we "knew" that, we might be able to respond to minimize damage or even gain an advantage in the "exchange." If we "knew" our customers were delighted with our product's reliability, we might do something entirely unexpected. We might run an ad campaign emphasizing our product longevity and make our competitor's "need" for a warranty extension a negative in the consumer's minds.

Modeling is rife with challenges, especially when all data are historical rather than experimental. We design experiments to control all critical factors and randomly manipulate a few factors to see how results of interest change. Good experimental design can often build a strong case for causal relationships between a few significant factors and results. Unfortunately, historical data is rarely so nice and neat.

Historical data is collected as it happens. There can be important factors that were not measured that "confound" the data. An example might be price and sales data collected when the economy was slipping into a recession. Unless the analysis considered macroeconomic factors, it could be useless. And finally, historical data is continually plagued by errors and missing data points.

The famous statistician, Dr. George Box, said, "All models are wrong, but some models are useful." Useful models are numerical approximations of reality that can predict essential factors from easily measured ones. They may be useful for a season and pass away into history or indicate very persistent relationships that always seem to be reasonably accurate. Hence, model builders should be neither too demanding of accuracy nor too confident of predictions. One of the arts in modeling is for the data scientist to understand the data, the relationships, and the data needs well enough to communicate to data users the reliability and limitations of any model they create.

Perhaps the sexiest concepts in data science today are AI and machine learning. They are closely related concepts that either focus on the process of "thinking" (AI) or more on the gathering and using the "thinking" (machine learning). The concepts bring together much of what we have discussed so far. Let's start with machine learning.

The most straightforward way to think about machine learning is to imagine an example – the Roomba. This amazing device uses machine learning to navigate your house and vacuum your floors. So how does it do it?

The Roomba has multiple sensors for detecting when it runs into something and multiple preprogrammed actions to continue vacuuming. It responds differently from a "dumb machine" because it remembers bumping into something and tries to map out a "better" performance next time. The machine learning part of Roomba is the sophisticated sensing equipment connected to large memory and a fast, powerful processing system able to recall and use gigabytes of data. Roomba can "learn" to avoid obstacles by rote but optimizing the time and efficiency of vacuuming requires a bit more. Designing efficient "thinking routines" ("algorithms") is the realm of AI.

To design a practical algorithm, we must know about the task. We must know the possible actions our machine can take, the ultimate goal of the process, and the criteria for selecting "optimal" actions. The choices and criteria can be objective and straightforward or nuanced and complex. Simple, probabilistic algorithms that choose between a few options have been very successful. For more complex and nuanced applications, neural networks seem to be more successful.

A simple algorithm might be, "If you can't go forward, then try going right, repeat until you are going forward again." A more sophisticated algorithm will look at many factors before deciding to go right. For example, additional factors could include the current position and previous choices at that position.

AI isn't very impressive if it is nothing more than remembering not to repeat mistakes. It becomes more helpful when algorithms successfully navigate "changes" to conditions. If the machine makes a "good" decision when conditions are different, we are more apt to say the machine has "learned." "Good" decisions generally follow specific decision patterns. We are creating a more powerful AI application if we "learn" these patterns, incorporate them into the decision algorithm(s), and then modify them with the newest data.

AI and machine learning sound easy enough until we investigate the details. It is one thing to allow Roomba to stumble and fumble around, learning how to vacuum a room. It is a whole different matter allowing a new AI system to replace a team of real estate "experts" for buying and selling houses (e.g., recent Zillow problems). The history of AI and machine learning is full of spectacular failures. Those failures are usually related to insufficient or irrelevant data leading to inappropriate conclusions. Nevertheless, there are many more, less spectacular successes with controlled experiments and high data quality.

An important principle derived from data science is the power of multivariate analysis. If we control our experiments and data quality, we can quickly and reliably evaluate multiple factors simultaneously. Furthermore, the available computing power makes multivariate analysis practical by solving complicated calculations and acquiring and storing mountains of data in milliseconds. These are frequently called "principal components" (from formulation studies), and getting to those components is called principal component analysis.

The power of this approach may not be immediately evident until one thinks back to our old chemistry or physics labs. You may remember the instructor yelling at the sloppy students, "Only change one variable at a time!" Years ago, if we changed multiple variables, we did not have the computing power to easily deconvolute the data to reveal which variables had the most critical impacts – especially when more than one variable had some effect. Today we can easily apply matrix algebra to deconvolute many variables under certain conditions (mostly related to linearity). With multivariate analysis, the DoE becomes a potent tool.

Example 17: Taking and Preserving Samples

A company attempting to sell a new technology into the corn ethanol market needed a reliable method for preserving samples pulled off the production line and sent to an outside laboratory for testing. The technique must stop the enzymatic reaction converting starch to sugars so that the "degree of depolymerization" could be accurately measured by the outside lab. Unfortunately, getting the samples to the laboratory in less than 48 h was challenging. Furthermore, taking samples from the production line was a costly interruption that the producer wanted to minimize. A literature study showed that sample pH, temperature, and storage time could be important factors. Testing each of these factors one at a time would have resulted in many dozens of samples. The experimental design reduced the number of samples to just eight by varying the key factors simultaneously and applying multivariate analysis to the data.

DoE is extremely powerful when:

- the data are collected under controlled conditions,
- the confidence is high that the relevant factors and results of interest are included in the data set in an unbiased way,
- the results of interest are known to be or can easily be shown to be linearly related to factors, and
- the measurement methods are adequate to differentiate "good" from "bad" results.

Most DoE software will quickly reveal the primary factors to consider, and the resulting model will indicate the relative importance of those primary factors. The factors of greatest impact will have the largest linear coefficients. In addition, the software will "report" potential "interaction" between factors in many cases. Finally, the software will suggest that some factors work in nonlinear combinations with each other.

The software will also generally reveal which factors are not relevant at the measurement precision and confidence levels achieved in the study and specified in the data analysis. Good software packages will also report on the suitability of the modeling using various statistics and graphical representations. And finally, the best software programs will assist in choosing the appropriate numbers of samples to be taken. They will also suggest the proper testing levels for various factors to achieve specified confidence in the model to be created. They will automatically randomize test runs and test sampling plans for "orthogonality." This last feature helps ensure that the various factor levels are randomized and each potential factor has an equal chance of affecting the multiple results. Without orthogonality, certain factors may be given an unequal weighting in the model and introduce unintended bias.

The general DoE approach can often be extended to broader systems with more work and increasingly more sophisticated modeling software. For example, linear models work for many nonlinear relationships and interactions if the data are "transformed." In addition, some software packages include nonlinear modeling and non-Gaussian statistics. These options, however, must be used with discretion. In general, they require additional testing precision and extra sampling points to verify that the models used are appropriate.

Historical data rarely meets any of the data criteria listed above. Historical data are usually "inherited" from other "studies." These studies had other objectives and controls not related to the issues. It is also doubtful that we will know the precision and accuracy of the measurement techniques. Historical sampling is rarely orthogonal. Frequently, the data set usually introduces a bias toward specific factors. Hence, it is uncertain if the data set fairly or consistently reflects random variations from trial to trial. Proceeding under these conditions is tricky, at best. We must test data to verify appropriateness whenever this is practical.

In general terms, the data set is "split" between a "learning set" and a "verification set." First, the "learning set" is used to create a multidimensional, linear, predictive model relating factors to results. Next, the "verification set" is used to "test" the model by replacing the "learning set" of data with the "verification set" of data. Finally, the predicted results are compared with actual results to see if predictions meet the acceptable statistical error.

The splitting of the sample set turns out to be a crucial step in the process and is the subject of much debate. The most straightforward approach is temporarily suspending judgment and treating the historical data set as a representative, random study. Thus, the "learning set" is simply a random subset of data taken from the whole data set. Generally, at least 25% of the remaining data are held back as the "verification set."

The researcher must select one or more parameters as results (outcomes), and the others are treated as factors potentially causing those results. These choices are not trivial. The researcher assumes causal relationships that may be incorrect and obscure or confound the real connections. Multivariate statistical software such as

Design Expert® by Stat-Ease (see [11]) can generate data fitting models. Potential principal factors are those that have nonzero model coefficients. The model coefficients indicate the potential relative importance of each factor.

The model can also show potential "interactions" between factors by looking at the form of the model (i.e., which factors or combination of factors make the best fit). The researcher can apply transformations and optional nonnormal distributions to improve model fit. The models generated can be powerful tools to elucidate potential relationships when the "verification set" predicts results within the model error. Nevertheless, the information from these models should only be considered reasonable hypotheses rather than statistically valid conclusions. The techniques applied by the researcher may be nothing more than sophisticated "curve fitting," a high-powered effort to "make" the data "prove" a prejudice.

We must never forget that the historical sample set may not represent our population of interest. Unless we can show that the historical data are an unbiased representation of our population of interest, conclusions drawn should be treated as no better than sound hypotheses yet to be proven. Whenever practical, researchers should collect new samples under controlled conditions. Using historical data to develop hypotheses and then verifying those hypotheses with a few new samples can be an efficient approach. Unfortunately, researchers too often rely solely on historical data. Researchers can get lost in the mathematical details and carried away by enthusiasm for "data science." It is the cause of much sorrow and woe.

> The ideal Technology Development process includes using extensive historical data to generate plausible hypotheses about key factors and their historical relationships with actual results. The hypotheses include estimated randomness in the data and "levels of interest" in results. Testing these hypotheses using new data generated under the conditions of a designed experimental protocol provides the statistical power needed to make correct decisions.

The most efficient way to develop new technology is to start with historical data and create targeted new data to verify models. This approach creates plausible hypotheses and eliminates biases, wasted efforts, unfounded conclusions, and inappropriate models quickly and efficiently. Researchers should follow this strategy whenever practical, especially when development costs are high.

Nevertheless, generating new data is often not practical for some factors. Unfortunately, new data is often unavailable when most needed – when the outcomes are the most significant. Examples include finding rare minerals of great value and treatments for some of the most lethal diseases. It can be impractical to run experiments with actual results (e.g., finding large mineral deposits or killing many patients).

When this is true, historical data may be the only viable approach. Researchers must comb the historical sample data carefully and objectively for
- potential bias,
- clustering, and
- measurement error.

This review should be completed and documented factor by factor because different factors will likely display different issues. For example, product preference data selected from the historical database of existing customers may not represent the product preferences of the market. Indeed, the historical customer base will be a subset of the general population that will almost certainly prefer the company's historic products over all possible choices (including those provided by competitors).

As noted before, no randomized analytical scheme can completely repair the validity of a biased dataset. Strategies to improve a historical dataset and make it more appropriate usually focus on cluster analysis. The idea is to find subsets of the historical dataset that are arguably more "representative" of the population of interest than the entire historical dataset. To do this requires additional "knowledge" of the dataset so that sampling is no longer completely random – it is "supervised sampling."

In supervised sampling, the researcher selects one or more factors to segregate sample points into subgroups more like the population of interest. For this strategy to work, the researcher must have verifiable knowledge of the population of interest and its relationship(s) to the historical dataset. The researcher attempts to reduce the impact of irrelevant or otherwise flawed, unrepresentative data elements in the historical dataset. This knowledge should rely on actual, extensive studies of closely related datasets such as those developed in the specific industry. For example, the theories of Pierre Guy summarized in *Sampling of Heterogeneous and Dynamic Material Systems* [16] are often used in the mining industry. For biological/ecological, supervised sampling, the models are often adaptive schemes like those presented by Thompson and Seber in *Adaptive Sampling* [17]. For more general systems, clustering and classification schemes are generated based on analysis of the historical dataset. Brunton and Kutz present several of these in *Data-Driven Science and Engineering* [18].

The danger, of course, is that the assumptions made about the system or the inherent bias of the historical dataset may be so flawed that the corrections made are insufficient, irrelevant, or, even worse, introduce additional biases. This last issue is perhaps the gravest danger to the honest researcher. It is not uncommon to use elegant modeling on flawed historical data to create embarrassing conclusions. It is all too easy to use powerful software to dismiss "inconvenient" data points and create a model that miraculously tells us what we hoped to learn from a dizzying array of irrelevant data.

Example 18: "Just How Dumb Are We?"

You can't do much Technology Development without having a story or two about doing something really dumb. I have several, but one of them was wildly entertaining. A good friend Jim and I were developing a two-catalyst system to turn syngas into ethanol. We had concocted a two-step process to make methanol using a standard commercial catalyst in the top of a reactor and then converting much of that methanol into ethanol in the lower section of the same reactor using a catalyst we had synthesized ourselves. We built a high-pressure reactor system with carefully designed temperature and pressure zones since the catalysts ran at very different conditions. We considered ourselves quite brilliant at the time.

We had tried starting up the system many times. Each time we got it running just fine, but it would have a sudden and dramatic drop in product yield after a short time. We reviewed our data very thoroughly and came up with all kinds of reasons for each failure. We would then make careful "improvements" and try again. We got better and better at getting the system up and going, but each time it failed pretty much the same way. It would start making lots of product and then almost completely quit. We finally noticed that the failure happened shortly after recycling the unused syngas back into the reactor to improve yield.

It turned out that our synthesized catalyst contained sulfur that bled a small amount of hydrogen sulfide into the recycled gas. We knew that hydrogen sulfide was a potent poison for the methanol catalyst, but in our enthusiastic "genius," we had somehow just forgotten that the small amount in the recycled gas would quickly deactivate the commercial methanol catalyst. The hydrogen sulfide killed methanol production and reduced ethanol yield dramatically. All our sophisticated reasoning about how to "improve" the system with each run turned out to be amateurish baloney. The plan never had a chance of working, and we should have known it. Somehow, we had convinced ourselves it should work and wasted weeks on a fool's errand. To this day, Jim and I laugh about when we realized just how dumb we could be. We try NOT to forget it.

We've been talking about the science of Technology Development. We have tried to focus on objective and mathematical pieces. In the next chapter, we will dive a lot more into areas of subjectivity, intuition, and "art."

Chapter 5
The Art of Technology Development

We mean by "art" the ability to create something of greater value and substantially different from the materials at hand, using techniques that are neither obvious nor certain of success. For example, we would not call "art" taking a soccer ball and making it green by spraying it with green paint. We have modified the color, but green rather than white and black is no real improvement. Furthermore, any hack can spray paint a soccer ball. If, on the other hand, we took a soccer ball, smudged it with some blood, drew a face on it, and gave it a character that caused us to morn when it was lost at sea, we might call that "art."

There is one other aspect unique about "art." It is "intuition." Many scientists, engineers, and educators have "pooh-poohed" intuition. Some claim it doesn't exist, while others vehemently oppose reference to intuition with a fervent, almost religious conviction. It is just as much a mistake to ignore intuition as it is to give it too much credence. Those who ignore it are desperately lost when confronted with the unknown. They are paralyzed when faced with an urgent need to decide without having all possible information. On the other hand, those who blindly "follow their gut" are doomed to make blunder after embarrassing blunder.

The "truth" is somewhere in between these extremes. We are frequently called upon to decide based on a "best guess." Some are better at guessing than others. They "see" weak relationships between factors sooner than others. Often, they have been in similar situations and vaguely recall something that is hardly a theory – it is more a hunch. Hence, persons with experience in the field often have the best hunches. When we are in a situation when a plausible hypothesis is needed, asking people with experience is often the place to find the "best guesses."

Example 19: The Mechanical Genius

My father was a high school graduate who did not get to go to college. Nevertheless, he had an uncanny ability to troubleshoot mechanical problems. He built roadsters and tuned them by ear. He invented several unique mechanical devices and held one patent assigned to his employer. His uncanny abilities earned him a job as the Superintendent of Maintenance at Union Wire Rope, a highly profitable subsidiary of Armco Steel. Making wire rope during the oil exploration boom of the late 1960s was about all that was keeping steel companies like Armco alive.

Union Wire Rope was a complex manufacturing plant that looked like a Jolly Green Giant version of a textile mill. Instead of bobbins of cotton thread a few inches in diameter, giant spools of steel wire 2 to 10 feet in diameter were whizzing around at speeds the eye couldn't follow. Acres and acres of these machines were drawing steel rods into wire, twisting those wires into strands, and then weaving the strands together to make wire rope up to 4 inches in diameter. These became the muscle of bridges, drill rigs, and draglines for giant mining equipment. These men made the sinews of America, and they knew it. Never mind that occasionally something would go wrong, and one of these spools might make two big holes in the 40-feet-high ceiling – one on the ascent

https://doi.org/10.1515/9783110451634-005

toward the heavens and another on the return to earth. It was all roar and clatter and dust and sweat. It was wonderful!

My dad was so good at what he did that upper management put him in charge of maintenance and engineering – something unheard of for a mere high school grad. More than that, they left him alone to "do his thing" and keep this monstrosity running. Hence, he frequently sneaked me into the plant during shutdowns. As a result, I got to see some of these wonderful machines at work.

Some stories told about the steel mills have an almost mythological flavor, but the one I'm about to tell is the Gospel Truth. I was there. I saw it with my own eyes because my dad had again sneaked me into the plant. The 4-inch closer was down. This was a big deal. It was the most profitable rope they made, and something about the 4-inch closer was not right. The rope had a kink in it, and no one could figure out why. So, they called my dad, and we rushed down to the plant.

Now making 4-inch wire rope in the 1960s was half technology and half art. It contains perhaps a hundred high carbon steel wires twisted together in smaller groups called strands that are then twisted together around a core to make the final rope. This last step is what the 4-inch closer did. The rope had to pass many quality tests, but the one that was the most telling was the simplest of all. A section about 10 feet long was cut and then rolled out onto the floor. If it snapped out straight as an arrow with a characteristic "thud," it almost always met every other quality test. It showed that all the wires were working together evenly. On the other hand, if it was kinked or shimmied or didn't "sound right," it almost always failed to meet specifications. The test showed something was wrong, but no one knew what more to do. The many adjustments made to rollers, stops, and tensioners had done nothing to fix the kink.

My dad and I rushed down to the plant. We were hustled through the guard shack and ushered into the inner sanctum of rope production. And there it was, eerily silent, the 4-inch closer. All eyes were on us. Some had that look of jealous disdain and others of feeble hope.

After a moment, my dad said, "Well, fire it up."

"Why? It's making crappy rope," fired back the Head Operator.

"I want to hear it run," quietly replied my dad.

"All right," snorted the indignant Operator, leaving out, for now, the customary punctuated ending, ". . . *asshole!*"

The Operator hit the green "Go" button. A snap of contacts, the buzz of a hundred amps, and the monster began to groan and move.

The 4-inch closer was a vertical machine with a couple of dozen spools of strand in a rotating cage about three stories high and perhaps fifteen feet in diameter. It didn't run at the breakneck speeds of the other machines, but it was massive. My dad watched it run for a few minutes while a crowd of anxious production people stood like they were in a diorama – whether mesmerized or frozen – I couldn't say.

After just a few minutes, my dad turned to the Production Foreman and said, "Replace the bearings on spool 2." Then he walked away. I was suddenly left alone with a double handful of production personnel looking at each other in stunned silence. Finally, the foreman scratched his head and said, "Well, you heard him . . . do it!" I skedaddled and caught up with dad. We drove home in silence.

Shortly after we got home, dad got a call from the plant saying the 4-inch closer was back up and running just fine. I couldn't wait any longer and asked how the heck he did that. He explained:

"Most production problems turn out to be maintenance problems. The production people often don't want to stop and fix stuff. I watched the closer run and noticed a strange sound that repeated each time spool 2 came around. I figured one or both of the bearings was bad and causing uneven tension on that strand. I wasn't sure which bearing it was or if that was the only problem, but in any case, uneven tension was not going to be good. It was a good guess, and replacing both bearings would cost almost nothing more than replacing one. The rest of it was for show. They don't have to know I'm not magic."

The Art of Human Relationships

By art, the Executive takes all the wrong people and molds them into an effective, functioning team. That may sound crazy, and for the genuinely rational Executive, it is not only crazy – it is impossible. We have been talking primarily about the logical, rational, scientific methods of Technology Development. Now we will talk about everything but rational behavior. We will talk about the unknown and the unknowable. We will be talking about people and relationships. We will be talking about motivating. We will be talking about "selling." Especially important will be "selling" to:

- Employees (in a broad sense)
- Funders (debt, equity, grants, etc.)
- Customers
- Stakeholders (including the public)

With Employees

The Executive can be sure that they will never have the "right" people for Technology Development. The "right" people are always scarce – for some situations, they do not exist. For the most part, people are not very good at Technology Development. Humans despise dealing with the unknown. We are the only animals we know of who will look up into the chaos of the stars and fabricate stories about how these animals, balances, and jars buzz around, determining the order of things here on earth. In general, we have such cognitive dissonance over the unknown that we easily deceive ourselves and imagine causality where there is only chaos.

Sadly, more education often does little to improve this. Instead, we learn to trade myth for math in our quest to wring out all uncertainty about the nature of things. There is hardly any contrarian observation that we cannot, through torturous diversions of insidious intent, turn into proof of our original preconceived notions. It seems as if the goal of education is to end learning and replace it with more subtle ways to delude ourselves.

We can combat self-delusion only by taking a consistently uncompromising stance against it. We do this by deliberately being skeptical. Furthermore, we insist on being most skeptical of our own ideas. Hence, when we set up our evaluation processes (discussed in Chapter 4), we make the hypothesis that we "want" our Null Hypothesis. We take as a given that it is not true and accept our hypothesis only if we can show, with high confidence, that the opposite is clearly wrong. Thus, we always take a position of bias against our desired outcome. We hope to squeeze out bias that can lead to delusion and fraud by such deliberate action.

This commitment to truth is one area where the Executive may find natural allies among the Technologists. Most Technologists, whether scientists or engineers, have a bias toward finding and telling the truth. Although this can be exasperating when applied to minutia, the Executive should tell everyone that the "truth" is not the enemy and that "truth-telling" is an enterprise virtue. Many Technologists will be stunned with disbelief from statements like that coming from a "business suit." Technologists frequently expect businesses to lie, cheat, and steal their way to obscene amounts of ill-gotten wealth. If the Executive can show a consistent commitment to the truth, they will frequently find considerable support for policies, procedures, and disciplines that creates a culture that discourages fraud. We will return to this theme in Chapter 6.

Of course, some of the personality traits of Technologists can get in the way of effective Technology Development. The Executive needs to be aware that Technologists often:

1. Are more loyal to their chosen field than to any organization,
2. Hold "regular people" in contempt and treat them shabbily – especially those in authority like the Executive,
3. Expect "peons" to do the dirty work while they think,
4. Are very competitive and even secretive when it comes to "their" ideas,
5. Have low self-esteem hidden beneath a veneer of bluster and aggression or extreme conflict avoidance,
6. Are from, or are currently participating, in dysfunctional family settings that make the forming of relationships very difficult,
7. Are not as competent in their current field they would like to be or even need to be, and often feel tremendous anxiety about their current position,
8. Are taught to be obnoxious, argumentative, and impolite by the institutions of higher learning – especially PhD programs,
9. Are petty, sensitive, and emotionally needy,
10. Are rarely motivated by money unless there is a perceived "fairness issue" (and then only negatively), and
11. Are genuinely frustrated by their lack of verbal skills and inability to communicate their ideas to others (even the "peons").

Please do not get me wrong here. I like Technologists. I often find them extremely interesting and even fun to be around. They just have quirky personalities that can be difficult. However, the Executive must have the ability to use this type of person to build the core of their Technology Development Team. Sadly, this group of misfits will have to interact with "normal" people from time to time. They will have to deal with lab technicians, purchasing agents, vendors, IT support, other executives, and in some cases, with pilot plant operators and their supervisors.

Many of the traits listed above apply to scientists and engineers alike. However, the Executive must consider one big difference between these two groups. Scientists generally like and even thrive in complex situations. They do not mind dealing with multiple, interacting factors. They seem to enjoy the mental challenge. They are often content to study the situation *ad nauseam*. Engineers, on the other hand, despise too many variables. They want simplified situations making choices easier and taking action more likely. They quickly get frustrated with thinking and prefer doing. Hence, there is always this tension – to study or to do – within the core of the Technology Development Team. Striking a balance within the core group of Technologists is rarely easy. The Executive must remain engaged to ensure that a balance is maintained.

The Executive can approach this dilemma in several ways. A few are listed below:

1. Do nothing and hope for the best.
2. Isolate the Technologists from others as much as possible.
3. Create competing teams of Technologists.
4. Retrain Technologists and those with whom they must interact.

The first three options assume that Technologists are incorrigible, and success is possible before animosity, excess costs, and organizational dysfunction destroy the project. These strategies work somewhat for high-value, time-critical, short-duration projects. Successful examples include the Manhattan Project and the development of the SR-71. Unfortunately, Technology Development is unavoidable in today's Knowledge Industry. Such "crisis mode" techniques cannot produce sustainable environments that create successful innovations year after year.

The "art" of the Executive is to "retrain" Technologists and those with whom they interact. First, the Executive must get the Technologist to look at their peculiar personality traits and see their strengths and weaknesses. This self-reflection is not as hard as it might seem. Technologists will analyze themselves if conditions are non-threatening. The key is to cultivate relationships that are open to discussion.

Next, the Executive needs to provide counseling and encouragement to those Technologists who would like to modify their behaviors. Of course, they will not suddenly become "people persons." But they can learn how their lack of patience, choice of words, unsolicited criticism, and other impolite behavior create unnecessary conflict, cost, and delay. Likewise, many non-Technologists should be made aware of the personality tendencies of Technologists. This knowledge helps them understand and tolerate the occasional "slip" into former habits more easily.

The above is not just a theory on how to manage Technologists. It works with many, though not all. Many Technologists realize that they have interpersonal challenges. Many have thanked me for pointing out the source of their problems and have worked diligently on improving their ability to deal with other people. As a result, many have enhanced their careers by learning to interact appropriately with others. Furthermore, many non-Technologists have improved their ability to cope with their Technologist co-workers once they understand the psychological side.

Three cautions are in order:

1. People rarely change their personalities. Do not expect people to become different. People do sometimes learn how to cope more effectively with their personality traits. You cannot expect more.

2. A significant number (I'm guessing about 25%) of Technologists make little effort to modify their behavior. A high percentage of these are the most senior, most powerful, and most important to the organization. Many are nothing much more than bullies.
3. Many HR departments often do not understand the personality traits of Technologists and are taken by surprise. Therefore, the Executive must have an excellent relationship with HR and provide additional training if they have not dealt with Technologists before.

If the reader gets the impression that the effective Technology Development Executive must become an amateur psychologist, this is not far from the truth. The Executive must become better at Technologist psychology than most professionals in the field. If done well, this can be a competitive differentiator for the organization. If done poorly, the organization will be like the many other technology-driven organizations that may make money but are terrible places to work.

One big struggle for the Executive is personnel organization. Technology Development functions change rapidly and constantly. These changes quickly render the status quo obsolete, and the Executive must "re-organize." Change is always painful. Often the "old guard" has little competency with the new challenge and cannot be expected to take a lead role in Technology Development. As a result, they have become "obsolete" and their future in jeopardy. Often Senior Technologists resist innovations.

This phenomenon is discussed in depth by Howard Gardner in his landmark book, *Creating Minds* [19]. Gardner reviews the lives and careers of seven famous modern creatives: Sigmund Freud, Albert Einstein, Pablo Picasso, Igor Stravinsky, T. S. Eliot, Martha Graham, and Mahatma Gandhi. He makes multiple conclusions about personality traits and career trajectories. Creative personalities vary from antisocial to sadistic. Tactics used by creatives run from self-promotional to extraordinarily self-promotional. They all had significant difficulty maintaining relationships with those closest to them (spouses, family, friends). Yet, they always seemed to have an extraordinarily loyal "servant" to handle conflicts and facilitate their self-promotion. Creatives are recognized "experts" in their fields during their younger years but make some "Faustian Bargain" with established authorities to maintain prominence in their chosen field or related fields after their creative prowess wanes.

Gardner asserts that these observations are the nature of things when dealing with most creatives. If he is correct, then the Executive **must** anticipate this situation and prepare well ahead of the various crises that are on the way. During the phase of creative prowess, the creative will have antagonistic and destructive relationships with peers, support staff, and management alike. They will tend to be intensely competitive with peers, dismissive and mean to support staff, and hold management in utter contempt. Eventually, their creative prowess will begin to fail, and they will blunder. They will support flawed technologies, and their egos will not allow them to see the problems or adjust their approach. Then, if they are lucky, they will make a Faustian Bargain to become a "senior scientist," a mentor, an ambassador, or some other person of prestige without significant responsibility for technical tasks.

One approach that the Executive might take is to use the creatives as long as the gain exceeds the pain and then throw them overboard when they become too much trouble. It would be naïve to believe that this approach is rare or unprofitable. Many technology-driven companies use this approach with great financial success while externalizing most of the cost of human carnage. There is a cost to the organization in pay scales, loss of wisdom, costs of retirement packages, turnover, productivity, wasted time resolving conflicts, etc. These are probably significant but hard to assess. Probably a greater burden is born by society in general from the "brain drain," loss of creativity, early retirements, cost of health care, cost of Technology Development, and general malcontent among the best, the brightest, and some of the most influential persons in our society.

A better approach is to create organizational flexibility around new projects and programs. For example, the Executive can establish ad hoc Project or Program Managers positions assigned to these new, yet-to-be-named projects. If the "old guard" is engaged in developing these ad hoc positions, they will feel much less threatened by them when it becomes necessary to create special groups to tackle new challenges. An especially effective technique is to give the "old guard" specific support and advisory roles to the ad hoc positions. In this way, they will have predefined roles and responsibilities that peers see as promotions rather than "putting out to pasture." In addition, they will have the ability to claim some success by their participation in the success of these new groups.

The Executive must curtail their natural inclination to "get the organization right." That is never going to be the case. Technology Development is much too fluid. Job Descriptions should be quite short and somewhat vague. Organizational Charts should remain buried in drawers. The organizational mantra should be:

1. We are a team with roles and responsibilities that adjust to the challenges at hand.
2. When executing a task, we will do our part as we have, together, defined it.
3. When discussing, thinking, and planning, we will listen to all opinions and ideas and judge them based on their merit and not the proponent's position on an Organizational Chart.

The Executive needs to balance egalitarianism and hierarchical command and control. The source of the "killer idea" is unpredictable. It can come from a Junior Technologist, the "Old Gray Hair" who sleeps through most meetings, or the janitor. The entire organization must understand when to act as peers and when to have strict command and control. Combined "tight" and "loose" control creates tension in any organization, but nowhere more glaringly than in Technology Development. The core players (scientists and engineers) have great internal struggles in this area.

They vacillate between wanting the anarchy of complete freedom of action and tyranny where things get done with great efficiency. The Executive must strike the proper balance. There must be enough structure that things can get done in a controlled and safe manner, and yet there must be the freedom to express new ideas and point out the "elephant in the room."

I have seen very effective ad hoc organizations. For example, a diverse group of Technologists designed a detailed "Managing of Experiments" program with minimal guidance by an external consultant and the Executive. The implementation of the program went smoothly because of broad "buy-in." Hence, when new projects came along, there was considerable peer pressure to grant the ad hoc Project Manager significant leeway in running the new Project with minimal interference from the "old guard."

If all this wasn't complicated enough, the Executive should keep in mind that the "right" people may change at various development stages of an organization. This problem is common when the same group tries to take an idea from conception to commercialization. The organization needs different skills for:

- Creating the basic idea
- Developing the idea to commercial viability
- Launching a business based on a new technology
- Growing and sustaining a commercial concern

Rarely does a small company develop a basic idea and take it to a growing business with the same people. Large companies don't even try this. Instead, they will pass each stage to a different group. Successful small companies will usually "stay in their lane" and work at only one or two of these stages, passing the baton to others via licensing, service contracts, or buy-outs.

We cannot leave this important topic of employee relations without listing some of the common types of "wrong people" for Technology Development. Unfortunately, these types show up so frequently that the Executive must deal with all of them at times. The Executive should be on the lookout for:

1. "Pollyanna" – No matter what happens, it is all good.
2. "No-Way" – There is never an idea that they can't prove won't work.
3. "Bully" – Right and wrong are determined through trial by combat
4. "Yes = No = Maybe" – They just can't decide what to do other than nothing.
5. "Just Want to Have Fun" – If it isn't easy and fun, it can't be done.
6. "The Know-It-All" – No one has a good idea but them.
7. "The Lazy" – They always want to collaborate but never really help.
8. "The Martyr" – They reluctantly volunteer for way more than they can do.
9. "The Loser" – Everything they touch somehow gets wrecked.
10. "The Handicapped" – There is always a reason why they can't do it.
11. "The Fraud" – They never did even half of what they claimed to have done.
12. "The Thief" – They have taken from everyone, and you're next.

Such folks should be isolated so that they can't do too much damage. They certainly should not be in positions of significant influence because they are "culture killers." They, by their nature, create conflict, disappointment, and failures.

A primary goal of the Executive is to develop a "culture" that drives Technology Development. Culture is the internalization of share principles that becomes the normative factor in group behavior. An organization has a culture when the enforcement of rules gives way to understanding "how things get around here." Few will applaud the development of corporate or organizational culture. The mix of freedom and flexibility required for Technology Development demands flexibility around rules. Modern culture focuses on rules (laws) that prohibit and not values that motivate. It is the difference between *malum prohibitum* and *malum in se*. Without a solid corporate culture, relaxation of work rules will often result in chaos and conflict – a corporate version of anarchy and lawsuits. The Executive must carefully and diligently develop a positive corporate culture to succeed long-term.

One final comment on culture. The Executive will be attempting to modify behaviors in ways that are not supported in American employment law or encouraged by our current culture, especially those bastions of mediocrity – our universities. For example, it is not the purview of the law to require that people get along, cooperate and are pleasant to one another, value the opinions of others, speak up when something goes wrong, or give praise and encouragement when they see something right. Furthermore, our universities compartmentalize learning so that creatives become technical experts while becoming woefully ignorant about business, organizations, psychology, relationships, and many other non-technical but equally essential topics. The PhD programs in the United States are especially adept at reinforcing dysfunctional personalities by using and abusing the untenured and students in an oppressive, creativity-killing system (see especially Berlinerblau's description of our university campuses in [7]).

If we want to make Technology Development work, we must work outside the current norms of dysfunction and loyalty to mediocracy. Hence, we must go beyond the everyday compliance with the law and outside our "safe zones" to achieve motivation by internal values. Only culture can instill and "police" the behavior required for creative success. If an Executive wants to have a Technology Development Team better than the competition, the organizational culture must naturally abhor the "acceptable" and encourage the "excellent."

Example 20: When Regulators Become Activists

I began my career in Environmental Chemistry. I worked at several different commercial laboratories starting in 1975. I often developed new analytical methods to test for toxic chemicals in complex sample types. We were developing new science in those days, and the EPA was at the forefront of scientific achievement. By 1990, however, the EPA had spent billions of dollars on massive cleanup projects, some of which had dubious value. The culture at the EPA had changed. It was no longer about science and truth. It had transformed into an activist organization driven by a strange mix of attorneys, tree-huggers, and career government bureaucrats. Robert Park recounts a

1990 debacle in Voodoo Science [10]. The EPA investigated the potential harm that EMF (electrical radiation) from powerlines might cause. The preliminary draft report that EPA leaked in May of 1990 contained multiple errors and vague statements. The report seemed to support the fantastic claim that power lines could cause cancer. It was up to the courts and several other agencies to quell the storm of panic and lawsuits. In 1997, three major, independent scientific studies from the National Academy of Sciences, Oak Ridge Associated Universities, and the National Cancer Institute finally squelched the EMF fraud. Sadly, some "baby boomers" still worry about their grandchildren living near powerlines.

We should not leave this topic without some essential guidance on motivational tactics. Many organizations attempt to incentivize Technologists (and others) with various performance-based monetary reward schemes. These are a waste of time and money. Money is not a strong motivator and, if tied to any performance-based scheme, it becomes a huge bone of contention. The scheme is never "fair" in the eyes of Technologists, and changing conditions almost always make it obsolete in a matter of weeks. Furthermore, Technologists are very adept at "working the system" to get benefits unrelated to actual performance. They are also very proficient at raising legitimate criticism to even the most thoughtful scheme. Such schemes result in endless discussions without any real improvement in performance.

For the reasons listed above, organizations should avoid performance-based bonus schemes based on subjective reviews. However, if the Executive is aggressive at culling out weak performers, those remaining will deserve some amount of recognition. In addition, when the Executive aggressively demands performance, tenure-based raises will reflect performance in the long run. And finally, percentage raises that reflect the company's overall success in the marketplace are straightforward, reasonable ways to reward long-term contributors for Team goals.

Hence, the only performance-based bonus systems that make any sense are some form of profit sharing. When the whole organization does well, the Team should share in the success. In some cases, it is possible to identify the financial performance group like a Project Team. When objective measures of economic success are practical, this can be a valuable motivator for the Team working on a large Project. More often, the best bonus systems reward increased overall company value. These bonus programs can be a wide variety of stock and ownership options, resulting in substantial rewards if the organization succeeds in its Technology Development. In addition, these broader-based motivational schemes tend to encourage teamwork rather than incessant moaning about individual plans' details.

With Funders and Potential Funders

We will deal with potential investors in more detail in Chapter 6, but now we would like to discuss specifically the "art" of managing the relationship with existing investors. The Executive must recognize that those who fund the Technology Development Project are humans with multiple motivations. Whether we are talking about the institutional investor's consultant, the state agency's Contracting Officer, the representative of the investor's CFO, or the visiting bank VP, the Executive will be dealing with people. The organizations that these folks represent will be looking at the wisdom of the investment. The people involved will be looking after their own best interests – not necessarily organizational interests. The Executive must be mindful of the many different interests at stake.

There will be a mix of issues motivating investors. These could include:
- Make money
- Feel good
- Save taxes

The Executive must learn why the organization invested or supported the Project and continuously speak to those issues. Failure to discuss with investor/funder representatives explaining why we are still on track to meet stated investor/funder goals is a formula for failure.

Nevertheless, only knowing those specifics is not enough. The Executive must speak directly to personal motives. The Executive must "sell" the people involved and convince them that the project is "still on track," regardless of the challenges. To do that, the Executive needs to know the key success issues for the representatives. The Executive must answer questions like:
- Why should this person believe me?
- Why would this person give their organization a "personal endorsement?"
- Why would this person go to bat for me if things are not going as planned/hoped?

When making these kinds of assessments, the Executive should consider Maslow's Hierarchy of Needs and apply them to the person representing the investor/funder. In many cases, the person will not be a decision-maker but, more often, a staffer. The long-term staffer frequently cares only about not losing their job. They may be more interested in making sure nothing "bad" happens than creating a great "success." Such a person needs constant assurance that everything is under control. They might want lots of thorough communication to get that assurance. Interestingly, a long-term staffer could be so risk-averse that they don't want to hear "bad news." They may want little communication unless it is "good news."

The long-term staffer could also be the "lonely heart." Of course, they don't want to lose their job, but they may have a strong desire to feel wanted or needed. They may desire involvement in even the spurious details to feel part of something important. They can be helpful, but there is always the danger that too much information leads to speculations about what might go wrong.

Occasionally a staffer is an "up-and-comer" desperate for a "big success." Such a person might relish the prospects of a big breakthrough even in the face of serious risks. Typically, this type of person is an owner or principal of a private investment organization. Such persons could even be "risk junkies" to the point that they become dangerous. They push schedules, ignore warning signs, dismiss issues, and generally encourage taking greater and greater risks.

The Executive needs to assess the persons in the communication and decision-making chains to effectively navigate the challenge of messaging. There is an interpersonal and organizational art to creating trust, likeability, and appropriate engagement. Unfortunately, Technicians are usually terrible at this. They generally don't like people enough to bother managing a relationship – often not even within their own family. Therefore, they should not represent the organization to outside parties.

The Executive rarely has the option of being the "face" to every investor/funder in every circumstance. Hence, they will need help. Furthermore, the investor/funders often want a "technical person" engaged in communication. This need creates both a dilemma and an opportunity. It is hard to find the "right" technical person to communicate with investors/funders. However, this can be a significant competitive advantage in getting and keeping funding sources. The Executive that invests time in grooming Technologists for positions that require interpersonal skills can reap enormous benefits for the organization down the road.

With Customers and Potential Customers

We will see later that every technology-driven company needs customer sales as early as practical. Hence, early customer relationship management is an essential facet of Technology Development. Customer organizations have objective goals when dealing the technology-driven company. Nevertheless, people operate customer organizations; they have personal preferences and goals. In most cases, the relationship between technology-driven companies and their customers is a many-to-many relationship. The technology-driven company has sales, executive, technical, and production personnel. The customer organization also has a mix of business and technical personnel. The interaction between the individuals forms a complex organizational relationship. The interactions and relationships formed are critical, especially in an environment full of unknowns like Technology Development.

Hence, the "sell" to a potential customer will likely be a "team sell," and the "servicing of the customer" will likely require cooperation between sales, technology, and production. The Executive must create a "team" that can effectively "sell" a product or service that meets customer needs, can be delivered on-spec, and can be supported effectively. The adequate support of the product or service will generally be a key success factor in ensuring follow-on sales.

Follow-on sales will be a critical factor in determining the organization's health. Technical selling (especially team selling) is a costly investment of resources. It is rare that the single, transactional sell can justify the cost of technical sell. Companies often lose money on early sales of new technology. Customers usually demand discounted prices to "pay" them for taking risks. Companies need long-term customer relationships to result in less expensive "re-orders." Therefore, the Executive must plan and train carefully to ensure that team selling creates long-term, profitable customer relationships.

The details of how to implement and execute effective technical team selling would fill another book. In any case, the Executive must deliberately create a multi-functional team where:
- Roles are clearly understood,
- Processes and steps are both practical and efficient,
- Flexibilities and constraints are managed "on the fly,"
- Sales seamlessly flow into production, delivery, and support,
- Deals strike a balance between revenue, quality, delivery, profitability, and liability, and
- The team functions as a real team and doesn't become a mob or a committee.

Creating an effective technical, team selling, and servicing organization is another "art" the falls to the Executive.

Example 21: The Team-Based Environmental Laboratory

The production laboratory is always a challenge. The workers in the laboratory are usually well-trained technicians (often university graduates) who must efficiently produce test results for a wide variety of customers with varying customer needs. The job can be tedious and frustrating for technicians who entered the sciences to do research.

The environment can be highly challenging for Environment Testing Laboratories where requests for services (tests) are not steady, types of samples vary, and customers' specifications for delivery, detection limits, and reporting styles can change. Mismatches between production and service requests can result in service disasters. One such Environmental Testing Laboratory had been entirely overwhelmed by sales personnel "overselling" the laboratory's capacity and capabilities. As a result, samples languished in storage for months and were often lost or spoiled.

The solution was implementing a team selling approach that put the outside salesperson with an inside salesperson. Both persons received sales training and at least rudimentary technical training. The outside salesperson had professional sales experience, and the inside salesperson had laboratory experience. The outside salesperson could not commit the laboratory to a project without the concurrence of the inside salesperson, who kept close tabs on laboratory capacity and capabilities. A larger specialized sales team that included operations and executive management handled the more extensive proposals and projects.

One laboratory pilot tested the process. After several months of perfecting the process, the laboratory implemented it at a dozen field offices across the country. It was expanded and improved, eventually becoming the primary method for managing the selling process from the smallest to the largest projects.

The HR Department expanded the program to a new career path by turning the inside sales position into a "Project Manager." The new career path relieved some of the frustration and "dead-ended-ness" of working in the laboratory.

With Stakeholders

Executives and Technologists tend to overlook the advantages of communicating regularly with "stakeholders." These are the individuals and organizations interested in the Technology Development Project. The reasons stakeholders become interested in a particular technology are varied and often subjective. Hence, the "interest" may not make any sense to the Executive or Technologist, nor might it be a "welcomed" interest. On the contrary, the "interest" may be strictly and unequivocally hostile and antagonistic.

Stakeholders include a wide variety of supporters and friends that would like to see the project succeed. They will often benefit from the success, although the stakeholder can't influence the outcome. The Executive may want to cultivate these relationships and keep stakeholders informed of progress. They may benefit the Executive later in both financial and non-financial ways. Occasionally, endorsement of a "third party" stakeholder with no financial stake in the technology can be a powerful persuader for potential allies and detractors.

Stakeholders also include regulators and the press. These folks are responsible for watching projects and reporting potential violations and hazards to others. Though it may not seem it at times, most regulators and reporters are just trying to do their job as they see it. The Executive can often help them do their job and promote their cause by making it easy for regulators and the press to get accurate information directly.

Of course, we would always prefer to communicate with those stakeholders who support us and help us achieve our goals. Nevertheless, it is hazardous to completely ignore any party expressing an opinion in this modern age of histrionics. Even apparently insignificant, uninformed, and irrelevant views can create a crisis, especially if handled dismissively or rudely. The manner and tenor of responding to an opinion can become THE issue regardless of its merits. Therefore, the Executive (or a designated representative) should answer every complaint or criticism politely and with relevant details.

Example 22: Flamed by a Blogger

A well-known green industry blogger harshly attacked a startup company working on a process to make renewable fuel additives. The blogger claimed that the yields for the process could not possibly justify the costs of making the additive. The blogger claimed to have expert knowledge on the topic and insisted that the approach was hopelessly flawed. Of course, management at the startup company was furious and wanted to strike back. Cooler heads prevailed, and a top executive wrote a polite response thanking the blogger for showing an interest in this breakthrough technology, acknowledging that the yield issues were crucial for success and that those issues were of the highest priority to the Technology Development Program. The blogger moved on to other "hot" topics of the day.

Some guidelines for the Executive in communicating with stakeholders:
- Determine who will do the communication and let others know the rules for communicating with outsiders.
- Always thank stakeholders for their interest – good or bad.
- Always be polite and NEVER attack at a personal level.
- NEVER imply that motives are nefarious or self-serving.
- Communicate often with supporters – infrequently with detractors.
- Don't promise what you are not sure you can or will do.
- Acknowledge challenges or blunders and state what you are doing to mitigate or remediate them.
- Use the opportunity to state what you are doing well, even when acknowledging something that has not gone well.
- Be objective. Be confident. Be upbeat.
- Refrain from venting regardless of how unfair the criticism is.

Having said all the above, dealing with self-proclaimed "stakeholders" can sometimes become a never-ending verbal exchange. Some folks have nothing better to do with their time than to eat up the time of others in useless arguments. The Executive may choose to ignore the rantings of an obvious crackpot or take legal action against someone whose claims pose a severe threat. There is no guarantee that either is the best action.

The Art of Matching Technology with Markets

Chapter 1 talked about the potential markets and how the technology needs to fit the chosen market. We also discussed the need to design technology features to compete in those target markets. Those features must yield benefits to the potential customer, and we must have "enough" data to make claims about those benefits at least plausible. The art comes in deciding which features will yield benefits and how reasonable or provable they need to be.

It is important to note that these decisions hinge primarily on the customer's perceptions. Even when dealing with B2B specifications that may be very exacting, the customer will have the last say when evaluating the plausibility of the claims made. Even government agencies who are required to follow strict rules (e.g., FDA) can exercise almost unlimited "veto power" if they are "uncomfortable." Hence, at some level, the Executive must make decisions about adding features and collecting data based on their "guesses" about customer perceptions.

> The Executive who is not familiar with the customers in the target market is at a grave disadvantage in setting priorities for the Technology Development Program.

A pivotal decision to be made by the Executive is when to cut off changes/improvements to the technology and "go to market." This decision can be more complicated and consequential depending on what "go to market" means. If that means building a $500 million plant based on the technology, the product produced must sell, or the project may become a half-billion-dollar debacle (e.g., Solyndra). On the other hand, if "go to market" means licensing the technology to a savvy producer, the risk is much reduced. Thus, we now come to yet another art form for the Executive.

The Art of Monetizing the Technology

Although we may congratulate ourselves on the clever models we develop for valuing technology, the real value is what someone will pay for it. Unless a Technology Development Team can turn it into cash flow, that idea has no monetary value. Therefore, the Executive must develop a sustainable cash flow strategy. We often call this "monetizing" the idea or the technology.

There are two basic ways to monetize an idea:

- Sell the idea to someone else or
- Use the idea to deliver competitive products/services.

These are two very different approaches. The first will require turning the idea into some form of intellectual property (IP) that can be sold or licensed. A common practice is to patent the idea. Patenting is an expensive, time-consuming process. The patent can provide exclusive use of the concept for 18 years, but there is no guarantee, even after receiving a patent. The idea must be economically important and reasonably difficult to substitute. If the idea is not economically significant, no one will pay for the exclusive use.

Moreover, if an idea is too important, it could attract the attention of major industrial players (and even the government) who will fight to disallow or circumvent the patent. If the idea is too easily substituted (even with slightly inferior performance), many "copy-cats" may appear. It could become a nightmare to defend the patent against a hoard of "knock-offs."

Developing and implementing an effective IP strategy is no simple task. It requires unique skills that are pertinent to the field of interest. It requires detailed knowledge of the industry, the competitive environment, the regulatory structure, and the history of patents. Some industries seem amenable to patent protections, while others ignore them. The IP strategy in the international oil and gas industry is quite different compared to the consumer "novelty item" industry. The Executive (or Executive Team) needs specific experience in IP in the field of interest. Without that, the IP strategy is likely to be a complete failure. Unfortunately, many inventors pursue IP strategies with little direct experience and fail. The inventors succeed only in providing detailed information to potential competitors, spending thousands of dollars, and wasting man-weeks of time.

As difficult as it might be for a fledgling company to manage an effective IP strategy, it is usually far more challenging to use new technology to make money directly. Much depends, of course, on the nature of the technology. Launching a new phone app is different from building a new, $500 million production plant around a new, "disruptive" industrial technology. Unfortunately, the innovator/inventor is frequently incapable of doing either. The innovator/inventor often has a specialized skill set that does not include the marketing, financing, and production savvy to take the innovation directly to the market.

Many ideas and companies fail at the point of monetization of the concept. Inventor/founder enthusiasm for the technology often blinds them to the need to think through this crucial step. When they do give it some thought, that same enthusiasm frequently clouds their judgment on the practicality of the monetization plan. Nowhere is this more common than when academics develop the technology. Academics rarely have the interest, skills, experience, or contacts to develop practical monetization plans. Technologies developed in National Laboratories and those funded by government grants (SBIR, STTR, USDA, etc.) fare only nominally better than academia in this regard. We recommend that every funding decision include a monetization plan with an independent review done by a firm experienced in the target market.

It is very common for funding agencies and investment groups to seek third-party technology reviews. Unfortunately, many of these reviews are unrealistic. Instead, agencies and investors tend to employ specific "due diligence" contractors who provide information in acceptable formats without relevant content.

Example 23: The Not So Expert Review

I will never forget being present for a third-party review of a technology funded by USDA and supported by a DOE National Laboratory. The "expert reviewer" was competent when asking specific questions related to laboratory analyses. However, the "expert" was utterly ignorant of the costs and challenges of building a plant to make the target product and hadn't even considered asking any important market-related questions – including the ability to meet market-required product specifications. The third party had a good reputation for reviewing projects, but not of this complexity or type. Hence, the review done was entirely useless for judging the probability of success for the project. The issues not reviewed – especially those related to the business case – caused the project to fail at a high cost to government funders and private investors.

It is hardly possible to overemphasize the need for a realistic technology monetization plan. The plan should make good business sense, and a competent third party should have objectively, if not skeptically, reviewed the plan. Unfortunately, the opinions of inventors and stakeholders are rarely objective. Inventors love their ideas without equivocation, and stakeholders love the possible outcomes. Even existing investors have limited abilities and resources to "research" a technology-driven deal,

understand it fully, and make completely objective decisions. Therefore, the Executive must engage the Technology Development Team and expert consultants to develop the plan to make money and move the organization through the long process of commercialization. This process will take many twists and turns often including:
- Asking for more money and changing the financial deal,
- Changing the technology and the Technology Development plan,
- Implementing the Technology Development plan,
- Reorganizing the company,
- Perhaps even changing the target markets.

The Executive can be sure that the Technology Development Plan will change. However, making changes will shake investor and stakeholder confidence. Therefore, the Executive must implement changes carefully and thoughtfully. Disciplined action instills confidence like nothing else can. Therefore, as we turn our attention to raising money for a technology-driven venture, we will frequently reference the discipline needed for success.

Chapter 6
Funding and the Discipline of Technology Development

To this point, we have been talking primarily about evaluation and planning for Technology Development. We have called this the science and art of Technology Development. We have tried to balance the objective and subjective aspects of the evaluation and planning process. We will now start talking about implementing a Technology Development plan. We will speak directly of disciplined, deliberate actions that allow course corrections without inciting panic.

Nevertheless, we should remember that reality has the nasty habit of disabusing us of our well-laid plans. We will certainly return to our plans, question them, and revise them from time to time. We will agonize over what we don't know, have overlooked, or have misinterpreted. In Technology Development, we can never be sure that we haven't made some colossal blunder and be prepared to take action when confronted with a crisis.

This chapter will discuss raising funds and the many disciplined actions required. It may sound strange to associate fundraising with disciplined action and implementation, but it is critical to note that planning, development, and funding are not standalone sequential steps. They develop together, they affect each other, and they iterate together. We must modify them simultaneously as we learn new information and find new challenges and opportunities. The novice believes that plans flow smoothly and logically into a profitable business. The truth is:

- Plans are always flawed.
- Funds are always inadequate.
- The startup is always on the brink of disaster.

This environment can be both exhilarating and cruel. It demands discipline, integrity, and courage throughout the organization. The pressures of such an environment can bring out the best in people. But, unfortunately, these same pressures can quickly turn the environment into a dystopia where self-deception becomes corporate fraud and epic financial disasters.

Example 24: When Scientists Go to Jail

In the early days of the EPA, one of the more difficult analytical challenges was testing dioxins at low levels in biological samples. Dioxin was one of the most toxic of all environmentally significant chemicals. It was a trace component in Agent Orange, a defoliant used in the Vietnam War, causing skin lesions and cancer. An explosion at a chemical plant making herbicides in Seveso, Italy, released enough dioxin to kill 3,300 animals and contaminated another 80,000 farm animals within a few days. There was a panic to develop trace level dioxin testing quickly. Three environmental

https://doi.org/10.1515/9783110451634-006

chemists developed unique analytical methods and launched a company providing rapid tests results for dioxin.

Initially, the company was quite successful. Unfortunately, the analysis for dioxin turned out to be more difficult, time-consuming, and costly than originally thought. There hadn't been a significant breakthrough in technology. As costs rose, backlogs of samples increased, customer satisfaction dropped, and profits waned, investors in the company put pressure on management to meet performance expectations. That pressure tempted middle management to cut corners. The pressure became too great somewhere along the line, and managers began fabricating quality control data. The fabrications grew into an elaborate fraud with fake computer-generated quality control data delivered to customers.

Eventually, government inspectors discovered the fraud. The laboratory was barred from all government contracting. One of the middle managers spent five years in federal prison. The company was sold to another entity for next to nothing. Many of the employees lost their jobs.

Example 24 is typical of how claims about technology can get out of hand. When added to financial pressures, there can be a toxic soup of excuses, exaggerations, rationalizations, and re-interpretations that can head down the path toward lies and fraud. More rules and regulations can rarely stop this slide. Once a culture embraces deception, there is almost no limit to the fraudulent creativity to circumvent regulations and quality controls. The elaborate, computer-generated schemes that the laboratory created were quite ingenious – strangely worthy of some admiration. What was needed was a type of discipline – not based on crime and punishment, but on shared values.

We are speaking here of discipline in a more classical sense. The ancients "disciplined" themselves and their novices by "exercising the disciplines" of their order, sect, or community. They practiced specific techniques to build their bodies, train their minds, and share ordinary virtues. Today we associate this kind of discipline with Eastern rather than Western thinking and find Eastern examples more instructive. An example of exercising discipline might be the *katas* of judo. These are a series of stylized demonstrations of various throwing, striking and grappling techniques. *Judoka* repeat these drills to perfection for numerous reasons, including aerobic exercise, strength building, technique building, and the demonstration of technical principles to students.

We occasionally find such practice in sports and the military in Western culture but rarely in other endeavors. Soccer players work eye–foot coordination drills with a Hackey Sack. Baseball players try to hit balls with a broomstick. Football players run across fields of tires. Marines crawl through mud while live bullets whiz over their heads. These activities are often called "training" and "drills." They build not only technical skills but also common culture, and yet, where do we see similar drills with mental, technical, or ethical skills? Where do we see them in business or academia? We do not. The result is general sloppiness in much of what we do – especially in how we think.

One task of the Executive is to drill their organization in the techniques – primarily mental and ethical – of Technology Development. An example would be using DoE on projects that may not need it. Another example would be sending employees to training classes and expecting them to implement and improve on what they learned immediately. Other examples would include writing reports, taking proper notes, completing, and reviewing laboratory notebooks, checking work, performing safety reviews, and discussing test results when they might not be strictly required. Employees must repeat these tasks to build the skills and habits to be ready when needed. People need to get accustomed to the "right way" of doing business at all times so that it is second nature when the "chips are down." Nowhere is this more important than ethics and when money – especially investor money – can be a strong corruptor.

It is a fiduciary duty of the Executive to constantly test and prove the ethical performance of the Technology Development Team. Nothing is more harmful to a Team's performance than a lack of honesty, integrity, or ethics. The Technology Development Team grapples with the unknown and perhaps even the unknowable. The rest of the organization and outside stakeholders depend upon the honest and thorough reporting of the Team on issues and ideas that they cannot evaluate first-hand. If the Team is dishonest or sloppy, others may pay a considerable price. Hence, even the slightest breach of ethics should be dealt with most severely. In most cases, the Executive must dismiss those who deliberately or frequently compromise the Team's integrity.

Example 24 Epilog: The "Zen" of Technology

The story of the environmental laboratory in Example 24 did not end at failure. Instead, a bigger laboratory group purchased the assets and restarted them. I was assigned to evaluate and rebuild the operation. I interviewed the employees and selected a small group of middle managers to be part of the rebuilding team. The team led a complete rebuilding program that reduced the staff to a core group and began a thorough retraining program. We stopped doing any dioxin analysis and expanded a smaller set of services. The training program focused on basic skills in this new market and included hands-on analysis of test samples to validate the training. After a few months of training and testing, we invited the government inspectors who found the data fraud to audit our laboratory. The skills and knowledge of our staff so impressed the auditors that they became advocates for our laboratory within government contracting circles. In about 18 months, the laboratory became profitable again. It became one of the premier environmental laboratories in government contracting within a few years.

Almost as important as ethical compliance is thoroughness in conformance to planned actions. The temptation to "cut corners" simply must be resisted. When dealing with the unknown, hardly anything is more confusing, confounding, and disheartening than trying to comprehend results from an out-of-control experiment. Many companies have gone broke chasing after encouraging but inaccurate information resulting from sloppy testing. Here the Executive will require the assistance of an HR department that understands the personality traits of Technologists. The HR Department must be aggressive in assisting the Executive with enforcement of fundamental labor law and the activities designed to build culture.

It is difficult to overestimate the need for this kind of culture to follow, assess, and improve a Technology Development Program. The temptation to parrot back the "desired" results or suppress "contrarian" results is powerful. When things are "going badly," the attraction can become irresistible to many – even to some of the most exemplary individuals. Robert Park describes the brain as a "belief engine" that can lead the best of us astray in Voodoo Science [10]:

> In humans, the ability to discern patterns is astonishingly general. Indeed, we are driven to seek patterns in everything our senses respond to . . . As we become more sophisticated, we seek out ever more subtle patterns. So intent are we on finding patterns, however, that we often insist on seeing them even when they aren't there, like constructing familiar shapes from Rorschach blots. The same brain that recognizes that tides are linked to phases of the moon may associate the positions from the stars with impending famine or victory in battle. That is again the belief engine at work.

Part of the Executive's duty is to protect the "minority opinion" from undue pressure to conform. Protecting the contrary opinion will be especially important when things seem to be in turmoil. When the pressure is on, the majority opinion is often the one that seems safe rather than the one that is probably right.

How Discipline Impacts Funding and the Growth of Wealth

Earlier, we noted how important Technology Development was to the generation of wealth and how the inefficient use of capital impedes the distribution of wealth, especially to the providers of innovative labor. The ideal situation is to use capital wisely so plausible ideas get an opportunity to become sustainable sources of wealth generation for investors, owners, employees, and government taxing authorities. Wise investing is patient investing. It is striking a balance between risk and reward that results in widespread economic growth and broad wealth distribution. The holders of capital and expertise (labor) are both rewarded. Both grow and become "renewable" resources for additional economic growth. Some indicators that this is happening are:

- An economically stable middle-class,
- Considerable economic mobility (especially opportunity for the poor to become middle class),
- Low poor/rich ratios (i.e., broad wealth distribution),
- Low interest rates,
- Ample capital (equity and debt) to fund small businesses, and
- Stable money (i.e., low inflation).

We should probably add that these indicators are just that – indicators of the sustainability of economic culture. Attempting to mandate these through government policy or regulation is a fool's errand. The draconian government controls employed to "fix" these indicators by fiat is like dunking a patient with a high fever in ice water. It may relieve some symptoms but does little to cure the cause. So instead, we propose that sound business judgment drives financing decisions. We now list some guidance in this area, realizing that some of our suggestions are novel. Later (Chapter 7), we will suggest significant changes to business financing.

The Biz Plan

I never cease to be amazed by the lack of practical, written Business Plans. Despite all the hoopla and the many books and articles, business plans are either nonexistent or worthless baloney. I think the problem is that those who talk, teach, or "sell" Business Plans spend their time talking about the format and never get around to a serious discussion of the content – especially any evidence that the content is valid. Hence, I will say very little about outlines, graphics, and format and focus on the essential topics, including purpose, content, and evidence. I will be taking two perspectives simultaneously – the writer and the reader. There is a purpose in writing a Business Plan. There is a purpose in reading a Business Plan. What are they?

From the writer's perspective, the purpose of writing a Business Plan include:

- Forcing the writer to consider the many issues involved in developing and monetizing a technology,
- Forcing the writer to consider, in money terms, the value of the technology,
- Documenting the current plan and providing a history of how it has evolved,
- Serving as a handy source of content for other communications to investors, consultants, employees, customers, and other stakeholders,
- Allowing third parties to assess the business proposition of the technology, and
- Serving as a measuring tool to track progress and scope drift.

The Business Plan is both an exercise and a tool for the writer. It is an exercise in that it requires serious thinking and some outside research. It should require the writer to perform a gap analysis – especially in areas outside the writer's "comfort zone." A good Business Plan includes the evidence that supports the claims made and identifies where evidence is weak or missing. Having no Business Plan is terrible but having one that is just propaganda is worse. Such a plan is a giant first step down the path of self-delusion, fraud, and spectacular failure.

From the reader's perspective, the purpose of reading a Business Plan includes:

- Assessing the thoroughness and sophistication of the management team,
- Assessing the credibility of the idea and the management team,

- Assessing the stage of development for both technical and business aspects of the technology under development,
- Assessing the potential for success (monetary or otherwise) by association with the Technology Development, and
- Assessing the most effective relationship(s) with the program.

The details required of these assessments depend on the reader's potential relationship with the entity attempting the Technology Development. For example, suppose the reader is a potential investor. In that case, the assessment will undoubtedly focus on possible monetary returns. Still, it will also involve assessing technology risk, business risk, and the investor's role in the organization. Crucial for an investor will be exit plans that could allow quick returns unhinged from long-term business risks. Under some circumstances (more on this later), long-term business risks will be of little concern to investors who want to quickly "flip" their risk to others. If, on the other hand, the reader is a potential employee, the long-term business risks could be far more important than a quick gain on a stock market deal. A deal that turns a quick buck for an early investor but doesn't create a stable business is not a good deal for a key employee.

The Business Plan should be as short as practical. Nevertheless, most Business Plans are far too short. They simply do not address many important issues. To make the Business Plan "readable," it should have a one- or two-page Executive Summary that includes straightforward statements ("claims") that include:

- A summary of the technology and the business need(s) served
- A summary of the IP strategy
- A summary of the monetization strategy
- A summary of the financing strategy
  - Debt
  - Grants
  - Equity including the "ask" and estimated IRR if known
- A rough timeline to:
  - First sales
  - Positive cash flow
  - Investor exits

The remainder of the Business Plan will be a multi-page document explaining the details of the claims made and presenting the evidence available supporting the claims. When the claims of the Executive Summary match with the evidence presented in the body of the Business Plan, the reader should be able to assess the credibility of the claim and the risk associated with it. The Business Plan should acknowledge significant missing, incomplete, or questionable evidence. If it does, it goes a long way toward meeting the requirements of "full-disclosure" to potential investors. A potential investor who receives a thorough and honest Business Plan like we are describing here has little ability to claim fraud later.

> A thorough and honest Business Plan presented to an investor, a potential employee, a banker, or a funding government agency can be an effective tool preventing later claims of fraud. Conversely, Business Plans that are little more than sales propaganda can become a millstone around the necks of those who promoted the plan if things go wrong.

Where practical, the body of the Business Plan should disclosure the assumptions, risks, and calculational methods used. However, the details of the calculations are frequently complicated to read and disrupt the flow of the narrative. Hence, the writer should add the detailed spreadsheets to appendixes in the back of the Business Plan where "experts" can review them. Showing your work and exposing it to scrutiny and expert review goes a long way toward heading off any later claims of fraud.

> In the US, it is not illegal to be wrong. It is, however, illegal to intentionally deceive for financial gain or cause harm through gross negligence. Thorough, honest, and competent Business Plans are essential.

By the way, Business Plans are never "finished." As new information becomes available, the old plan becomes obsolete and must be revised. Of course, the Executive must control revisions and document changes. It must remain clear what is the current, official plan so that everyday activities align with current priorities. It must be possible to go back into history and know the official plan at any time. Decisions aligned with promulgated objectives are defensible decisions that are unlikely to be criticized later.

Funding in Large Organizations

Technology Development is not always an enterprise activity, nor is it always seeking outside funding. Technology development frequently occurs inside an existing organization and may be only a small fraction of its activities. These organizations may have specific budgets for approved research activities. A "Business Plan" for Technology Development for these organizations seems odd. Nevertheless, the best of these organizations have policies and procedures to create formal plans to manage research and development activities. Those plans may be called Research Plans or Development Plans but serve a similar purpose and cover a similar scope as the Business Plans described above.

An essential element in any R&D Plan is to align technology development activities and business realities. Technology that the organization cannot monetize is a terrible investment. As soon as it becomes evident that the current strategy for monetizing a technology is unlikely to succeed, the Executive should stop investing in that technology. Furthermore, the R&D planning process must respond to the ever-changing business environment. Failure to meet emerging business challenges with practical Technology Development efforts can become an existential threat to the organization. Hence, all R&D/Technology Development planning processes MUST have recurring input from the company's Business Intelligence sectors, especially those that retain contact with key customers.

It is difficult for large organizations to maintain effective R&D planning processes. The processes frequently become ossified around chasing the irrelevant pet projects of upper management and making the environment miserable with endless meetings, reviews, and reports. As a result, many large companies have eliminated their R&D groups and have decided to "buy" the technologies that others develop. They have decided to focus their attention on taking technologies to market rather than developing them. This "downstream" strategy can be effective for some established manufacturers – especially where quality standards and manufacturing costs are high (automobiles, drugs, medical devices, etc.). The danger, of course, is that they frequently miss early markets and disruptive technological changes. They can also become devout "technology killers," using their market position to create barriers to new technologies. These technology killers can become the enemies of wealth growth and, more significantly, the distribution of wealth.

Not all large organizations give up trying to do their own Technology Development. With some regularity, large organizations implement unorthodox methods to encourage creativity. To be sure, things can get out of hand, leading to the unauthorized use of equipment and personnel, but there are amazing stories of creative genius in some of the most unlikely places. Hardly any account tops the Lockheed Skunk Works:

Example 25: Kelly Johnson and the Lockheed Skunk Works

Kelly Johnson, the son of an immigrant family, became the most celebrated aircraft designer in US history. The notable chapter of the story was the rapid development of the SR71 Blackbird spy plane. This almost mythological beast flew so high and fast that it was invulnerable to the most sophisticated surface-to-air missiles that the Soviets could muster. This was no idle brag. Various ground crews shot about 4000 missiles at SR71's without a single hit. It really could fly faster than a speeding bullet – up to 3.5 times the speed of sound.

It is an "urban myth," however, that the Skunk Works was some rag-tag bunch of bohemian "air-cowboys" following a "maverick" leader. On the contrary, the Skunk Works had a long history of developments going back to the 1930s. Johnson was a great engineer. He won his first prize for an aircraft design at age 13. He made his first contribution to commercial aviation while still at the University of Michigan. But his real "genius" was in motivating and organizing other engineers in highly effective teams. He managed small, agile teams and challenged them to meet rigid specifications and tight timetables. He insisted on strict budgeting and thorough documentation but minimized unnecessary paperwork. He insisted on having too few of the very best and the highest-paid engineers in the industry.

The Skunk Works created the U2 spy plane during the early 1950s. Nevertheless, Johnson realized that the U2 and similar planes were becoming increasingly vulnerable to improved Soviet surface-to-air missiles and interceptors. So the Skunk Works began working on a high-speed U2 replacement almost two years before the famous Gary Francis Powers U2 shoot-down. When the US government wanted a quick alternative to the U2, the "miraculous" Skunk Works response was, in part, Kelly Johnson just staying up with the "industry" by paying attention to customer needs.

Keeping a Technology Development operation funded in large organizations is challenging. It is not so much the cost as the cost/benefit ratio. Large organizations often have the money to support Technology Development, but other uses of those funds such as operations, marketing, and acquisitions may pay better dividends to shareholders. There is always internal competition for funding, and the Executive in charge of Technology Development must be able to "sell" the financial benefits to shareholder representatives. Here the Kelly Johnson story is most instructive.

Johnson had 14 "rules" that he enforced. These were not the fuzzy "feel-goods" suitable for publishing in business management books. These rules were hard-nosed, specific guidelines for cost-effective government contracting in the aerospace industry. They included detailed guidance on keeping workgroups small, organizations "flat," budget reviews up-to-date, drawings simple and easily retrieved, specifications clear, communications with government agencies focused, reports to customers on-time and straightforward, and keeping secrets secret. He is credited with the phrase "KISS – Keep It Simple Stupid." Johnson's rules kept success rates high, overheads low, and the best engineers working on the next big customer "need" before the customer knew they "needed" it. Johnson was always able to justify previous "investments" in the Skunk Works financially and "sell" how continued "investment" would payout.

Kelly Johnson started life as the dirt-poor son of Swedish immigrants in a backwater Michigan mining town. His father ran a small construction company, and his mother washed clothes to make some extra money. Kelly would sometimes help by delivering washed clothes out of his wagon. He worked hard and got a decent education in aeronautical engineering. Johnson went to work for a good company and used his innovative labor ("human capital") to bring himself out of poverty and help his many colleagues build their wealth. Kelly treated the Skunk Works like a for-profit business, and he was the Executive Entrepreneur. He and his team fought for contracts, won them, and delivered them on time and on budget. The labor Kelly and his colleagues invested in Lockheed continues to build wealth for the employees, owners, subcontractors, and suppliers of Lockheed-Martin Corporation, as well as the USA and its allies.

Kelly Johnson was a brilliant engineer, but nothing he did was magic. He instinctively applied basic rules of business management to engineering. He integrated the multiple facets needed to be successful into specific principles relevant to his field. It began with assessing market needs, translating these into design specifications, identifying required innovations in materials and manufacturing processes, evaluating and testing potential solutions for performance, reliability, and costs, and continuously improving final designs. This approach would now be called Systems Engineering. We now see this approach in various fields where innovation is needed, processes are complex, development requires many cooperating disciplines, and the cost of failure is high.

An excellent example of how Systems Engineering works for designing and manufacturing medical devices appears in Martin Coe's book *Robust Systems Engineering for Medical Device Design* [20]. Coe follows the process from customer need to profitable production in the high-cost, high-reward medical device industry. In addition, the book illustrates how Systems Engineering can help manage the interplay between complex engineering risks and business requirements in industries far from the aerospace and defense industries.

Funding in Small Organizations

Unique opportunities and challenges arise when funding and managing Technology Development in small organizations. The consistent questions are:
- Do we have the resources we need?
- What are we going to do about the resources we don't have?

For the small organization, the answer to the first question is always "no." However, answering the second question is where the work begins. What resources don't we have that we must have, and will we get them?

Everything that we have talked about throughout this book must get done one way or another. You must have:
- The right personnel,
- The right equipment, and
- Sufficient funding.

In a small business, the Entrepreneur is usually responsible for finding the right people, equipment, and funding necessary to develop the technology and make it valuable. Unfortunately, the Entrepreneur is often not capable of doing this. Far too often, they are experts in a limited number of aspects, but their ego is too big to allow them to delegate important decisions to others. Entrepreneurs are also often poor judges of character. Too often, they "project" their personality onto others and expect employees, consultants, bankers, investors, and customers to be like them. As a result, they are not only disappointed, but they are also frustrated and angry when others don't work as hard, do not take as much risk, do not share the same vision, or do not have the same talents as they do. Successful Entrepreneurs temper their egos and find others to help them succeed at Technology Development.

Similar comments apply to many neophyte investors. See, for example, the chapter, "The Angel Investor Learning Curve," in Winning Angels, by Amis and Stevenson [14]

Finding the right people, equipment, and funding should be considered a type of "make or buy" decision. The Entrepreneur may wish to have them internal to the organization, but a better choice is often to bring in partners or investors of different types. Nevertheless, bringing in the wrong people or bringing them in at the wrong time can destroy chances for success.

The Entrepreneur, Inventor, or Founder generally needs "expert" help in identifying knowledge "gaps." Hiring a business consultant with experience in Technology Development (including funding) can go a long way toward avoiding being blindsided by issues. Likewise, investors ought to consider hiring an outside consultant with relevant experience. Of course, it is also wise to pay attention to what expert consultants have to say. It is surprising how often Executives ignore good advice when it contradicts the existing "narrative."

Example 26: Denial – A River That Drowns Many an Ill-Conceived Scheme

Even the best and brightest organizations struggle with collective ego. I was a member of an Executive Team responsible for merging two large organizations. These organizations had been competitors for many years and decided that merging operations and cutting redundant costs would improve their bottom lines. The company putting up the money decided to send the Executive Team to a well-respected merger and acquisition consulting group to get some training on making the merger a success.

The training was fascinating, and the consulting group gave multiple examples of similar mergers that they had shepherded. The consultants made it very clear that in every merger they had ever done, the merged company's revenue was 10% to 15% lower than the sum of the revenue budgets of the two independent firms. There were at least two reasons for this. First, the independent revenue projections of the two firms counted on winning some of the same work. Second, a review of the budgets always revealed some "double counting" of the same work.

Furthermore, a close review of customer opinions indicated that each group had work that had come as "disgruntled" customers of the other group. A merger resulted in some of those customers refusing to deal with the merged company. As a result, the CEO of the consulting group warned against expecting to keep all of the anticipated work. Instead, we should aim for about 85% of the combined revenue budget. We should consider anything above 85% as a very successful merger.

This advice seemed reasonable. I was glad that we had been given the "straight scoop" on what to expect. I didn't expect the "meeting" that would "break out" in the elevator headed out of the building. No sooner had the elevator door closed than our CEO announced that there would be no "dip" in revenue. Instead, we were going to be "different." We were going to get 100% of the combined revenue budget.

I went back to the office and poured over the combined revenue budget. Sure enough, there was about a 15% "overlap." Keeping all the anticipated revenue was a mathematical impossibility. However, the good news for me was that accounting had miscalculated the budget. They had blundered on my portion of the revenue. I was going to exceed my assigned budget almost no matter what I did, and my counterpart with the rest of the budget was going to fail miserably. I contacted my counterpart and let him know what the math was saying. At first, he didn't believe me, but it didn't take long for the math to begin to "work." He was always falling short, and I was always looking like a genius.

Fortunately, our CEO was a great guy, and when we both went to him to show him the data, he immediately "got it" and quit beating up on my counterpart. As a result, we are still friends today. And, by the way, we worked well together and managed to keep about 90% of the combined revenue budget – better than most mergers.

Technology Development Startups – A Peculiar Subset of Small Business

Whereas small businesses pose some significant challenges for funding, the startup is an incredibly challenging case. The issue is a lack of revenue. Investors hate to be "pre-money." Too many bogus valuation schemes (see [14]) depend on multiplying revenue by something. Most of these schemes are useless without historical revenue, and investors must SWAG numbers. They loathe to do this, and for a good reason.

At least 80% of new businesses fail in the first five years. This SBA statistic is surprisingly consistent year upon year. However, I think this may be a bit optimistic for companies trying to do Technology Development. In other words, I believe the relevant statistic may be worse because Technology Development startups have some peculiar challenges. Among these are:

- Little access to bank financing
- Difficulty in attracting qualified executives (CEO, CFO, CTO)
- Exacerbated "Valley of Death" (early stage, negative cash flow)
- Multiplied Risks – One serious blunder can destroy the business
- The Entrepreneur/Inventor/Founder usually become a problem

Because the business is just starting up, there is no cash flow to tap. Banks cannot lend money without having at least some "excuse." Revenue and banking history are the most common "excuses" required. Banking regulations have changed since the financial crises of 2007/08, making "neighborhood banking" and "reputation banking" difficult if not non-existent. Even programs designed to support rural and minority business needs are challenging to tap unless there is some revenue history. Hence, the highest priority of any startup business is to get to some level of "revenue." This observation bears repeating for emphasis:

> The highest priority of any startup is to get to the point where they have revenue.

Because startups have no history, they cannot attract qualified C-Level (Corporate Level) personnel. Many talented persons are not interested in taking risks or investing time in a startup. Furthermore, those who seem to be interested are frequently unsuited for small businesses. Many of the available C-Level persons were either dumped out of old companies because of poor performance or were "damaged goods." They often have personality flaws that make them poor managers or worse – sometimes MUCH worse.

In most cases, C-Level personnel who have been working in large, technology-driven companies have no idea what it takes to run a small business. Extensive staff personnel insulated them from many of the personnel, regulatory, organizational, budgeting, and operational challenges that small business owners must master. As a result, experienced C-Level managers frequently become overwhelmed and simply "shut down" under stress.

Because Technology Development can take time and investment, it is frequently associated with a period of negative cash flow, ominously called the "Valley of Death." Initial idea conceptualization can be relatively cheap. Initial ideas often can be "free" because a government agency developed them under contract. However, ideas developed under government contracts are rarely ready for commercialization. Developing all the pieces and collecting the evidence necessary to demonstrate a business case for new technology can run into the tens of millions of dollars and many years. It can require specialized equipment, years of laboratory testing, and pilot work costing hundreds of millions. Funding Technology Development Programs at this scale is well beyond the reach of most "family and friend financing." It requires equity capital which is a sophisticated financial "sell."

The multiple aspects of a sophisticated Technology Development Program multiply the risk of failure. Just as the failure of one critical component can crash a space shuttle, the failure of a single component can crash a Technology Development Program. This critical component failure can be from a missed technical detail or the wrong person at a crucial job. A 5% failure rate of a component multiplied by hundreds of independent decisions can reduce the probability of success to less than 1%! Startups must make hundreds of correct decisions with VERY few bad ones to complete a Technology Development Program.

And finally, the original Entrepreneur/Inventor/Founder can become a negative influence. Changes to the technology-driven business realities are almost certain to occur during the Technology Development Program. Often the original Entrepreneur/Inventor/Founder becomes disillusioned and even hostile to these changes. It is not what they wanted to do – the evidence and the business case be damned. With few exceptions, the idea's originator will not be around to see the commercialization of the technology.

The Funding Process

There are at least two ways to look at the funding process. The first of these is a historical and chronological perspective. Thus, we look at the typical sources used throughout the development process. The second conceptual approach attempts to identify the most likely sources of outside, third-party capital. These outside sources supplement personal, friends, and family funds that are usually inadequate for most Technology Development programs.

Specific funding sources tend to follow the history of a Technology Development Program. The earliest sources are from the Entrepreneur/Inventor/Founder and those closely associated with them. The transition from one source to another is often both fitful and awkward. The principles presented in this book and summarized below will help make those transitions less sporadic and more productive.

The sources of funding (generally we call this "Capital") are listed below in more or less the chronological order they usually occur in a Technology Development Program. They are:

- Personal Funds and "Sweat Equity"
- Family and Friends
- Government Grants, Loan Guarantees, and Research Contracts
- Cash Flow from the Business
- Bank Loans
- Angel Investors
- Crowdfunding
- Strategic Partners
- Venture Capitalists
- Institutional Investors

With few exceptions, the first thing that happens in Technology Development is that an entrepreneur has an idea that they think should work. They "believe" in it, often for reasons that they cannot fully articulate. It is probably more than a "hunch." It probably has some evidential basis, although it may not be convincing to anyone other than the Entrepreneur and close associates of the Entrepreneur. The Entrepreneur decides to become an Inventor/Advocate for the idea and starts investing time and personal resources in the concept. The Entrepreneur often brings along others who will put in the time, expertise, and limited funds to help with the, often informal, "understanding" that this "buys" them some interest in the idea.

We often call these early investments "sweat equity" because it is primarily an investment of labor. These early investments could become a formal sweat equity agreement at some point. Unfortunately, most of these agreements become problematic later since the Entrepreneur and his closest associates usually fail to make clear agreements. Frequently the "agreements" give away too much equity to some or all the participants based on completely unrealistic expectations and valuations. Disputes at this early stage can destroy the value of the technology by making it impossible to attract outside investment later.

> Sweat equity can be an effective method for getting Technology Development started. Entrepreneurs and their close associates would be wise to use a consultant experienced in the area to develop reasonable agreements. Attorneys are the last persons you would want to use. They can write a legally binding contract but cannot construct a fair deal. Attorneys create and manage adversarial relationships. Get the agreement worked out first, and then have an attorney

> write it up. You may find that you cannot agree. That's good to know upfront before investing time in a "deal" that won't work.

Either contemporaneous with or shortly after launching a Technology Development effort, the Entrepreneur will often tap trusted colleagues, family members, and friends to help finance the effort. These are personal loans of money, space, or equipment in many cases. Sometimes it is entirely gratis. Sometimes it has attached vague notions about a future benefit or at least return of the assets. Therefore, it is essential to clarify, in writing, expectations to avoid future conflict and legal entanglements. The Entrepreneur is often reluctant to formalize relationships, but this is a big mistake. Unwritten relationships frequently lead to misunderstandings. That is bad, but not the worst that could happen. One party may attempt to benefit from a misunderstanding unfairly. Hence, a "misunderstanding" could be a cleverly disguised plan to defraud, manipulate, and steal.

At this very early stage, loans are the best instruments, and interest or rent is the best way to "pay" for the use of the assets – whether money, property, or goods. It is generally a bad idea to develop schemes based on future ownership (e.g., "convertibles") since so little is known about the value of the technology.

> Entrepreneurs should avoid trading off equity (i.e., ownership) as much as practical – especially early in the development process when business and IP values are low. It takes large percentages of equity to raise small amounts of money. If the business or IP is successful, the portion traded off will be costly financing.

Other valuable instruments for financing at this stage are government grants, loan guarantees, and development contracts (SBIRs, STTRs, CRADAs, etc.). These can be low-cost instruments but can have complicating features. With few exceptions, government programs (federal, state, and local) focus on policy issues. The grant or loan will have specific focus areas such as rural development, minority business support, economic opportunity/revitalization, education, and a host of other primary policy focus. The policy goals may not align well with the business needs of the Technology Development. An example might be a beneficial new technology to improve oil exploration. That would have almost no chance of receiving government funding in an era when the government is hoping to promote developments in renewable energy.

It is a mistake to believe that government contracting can be "free money" for Technology Development. Three factors come into play that can make government programs ineffective:

1. Many programs are set-aside for non-profits and academic institutions,
2. Programs that are available to small businesses are often not available to start-ups, and
3. Some programs require a commercialization "roadmap" that can be difficult for small businesses to create.

The first problem is a relatively new phenomenon. We will look at non-profits and academic institutions in more detail later. For-profit companies (our primary focus) are not eligible for government work set aside for non-profits and educational institutions. The size, scope, and frequency of set-aside government work continue to grow. Even worse, many programs now require for-profits to "team" with non-profit or academic institutions. As we have mentioned elsewhere, the value of a technology comes from solving a problem that creates a cash flow. Without a cash flow analysis, the technology is worthless or priceless, depending on many subjective factors. Unfortunately, subjective factors disconnected from markets defy objective analysis and rational planning. Collaboration with non-profits and academic institutions can thoroughly complicate setting priorities, making the development activities inefficient and possibly useless for commercial purposes (see Example 3 earlier).

The second item can become a barrier to startups engaged in Technology Development. Many programs specifically bar startups and many more "discourage" them through the selection criteria. This tendency is too bad since we are likely to find some of the most promising disruptive technology well outside existing companies focused on their core technologies. If we had reliable methods to evaluate technology, we could relax many irrelevant selection criteria.

Companies and investors can easily overcome the third item by hiring expert personnel or consultants with specific experience in winning and managing government contracts. Government contracts require effective contract management and reporting that could challenge company resources and distract from important Technology Development priorities. Government agencies will tell you otherwise, but they don't understand the complexities of business. Therefore, companies must have expertise in this area, or they will lose contracts that they should have won or find themselves trying to perform on contracts that don't help with their Technology Development. Unfortunately, too many business managers, investors, and government contracting officers have unrealistic views of their abilities and backgrounds and blunder into trouble.

As the Entrepreneur moves from personal toward third-party financing, they will want to use debt financing as soon as practical. Debt financing is usually the least expensive way to finance a business – perhaps even less costly than government grants and contracts when considering the hidden costs of working with the government. Debt financing is the first type of financing where hard evidence of repayment will undoubtedly be required. The business planning that we have discussed throughout this book will become instrumental at this stage and subsequent stages of financing.

Bank financing is the most desirable type of debt financing because it usually has the fewest "strings attached." Unfortunately, the business is unlikely to receive standard bank financing without a revenue stream. Banks naturally want to be paid back on a reasonable schedule, and payments from ongoing business revenue are simple solutions. Hence, every startup business must place a high priority on developing a continuous revenue stream. Once the company has established a revenue stream, a lender can collateralize the loan from future earnings.

A hybrid approach that can bridge to bank financing is the government-guaranteed loan. These will have some strings attached. Frequently, the money must be used for specific purposes or deployed in particular areas. Although some government agencies support pre-revenue loans, this is the exception rather than the rule. Most programs now do little more than reduce bank risk for disadvantaged groups or locations. Examples include loan guarantees for minority businesses, rural businesses, "green" businesses, etc.

We should not ignore the most common ways to fund a startup business. Many entrepreneurs start very small and support the early stages from:

- Personal savings,
- Earnings from their "day job," and
- Profits from their fledgling business.

Many businesses start in the basement or the garage. Examples include giants like Coca-Cola, Hewlett Packard, Apple, and Google. Anyone going this route should know that it is challenging work and that few succeed. Entrepreneurs usually begin with what they love to do. It is more or less an avid hobby. Eventually, the entrepreneur must decide to make the hobby a business and provide customers with what they want and like. Gardner calls this a "Faustian Bargain" [19]. Whether a new photography technique or a new gadget, the entrepreneur must focus on what others want rather than their peculiar desires. Too often, the "hobby" never becomes a business. The "innovator" turns out to be a technologist – "married" to an idea – and not an entrepreneur that adjusts to customer needs.

Example 27: A "Hobby" Grows into an Internationally Known Business

A gun enthusiast that grew up in rural Missouri went to work at a small gun repair shop in a bigger city. The owner died, and the gun enthusiast (now a junior gunsmith) teamed up with a friend to buy the business using personal savings. The shop was too small to make it a standalone company. The gun enthusiast moved the equipment to the basement of his house and launched a small operation working on rifles and handguns for himself and his close friends. He got a "day job" as an electrician in a steel mill. The pay was good, and the hours were reasonable enough to give him the money and time to buy more equipment and slowly build his "dream business."

His big break came when he became well known in Practical Police Combat shooting competitions. He augmented his ability to "accurize" police revolvers by patenting and selling an adjustable sight. He spends years supplying competitive shooters with competition-grade handguns at a good profit. Over time, he built a house and a shop where he would eventually move his business.

The "Faustian Bargain" came at the encouragement of his son, who eventually joined the business. Rather than focus on labor-intensive gunsmithing work, the company began to sell to other gunsmiths and general gun enthusiasts, jigs, fixtures, and gun parts (some patented) for a wider variety of guns. The company began to use more contractors to make some of the fixtures and parts rather than "custom making" everything. They also began to sell through dealers and on the Internet. The company had become a highly profitable business rather than a hobby.

Cash from a good "day job" and the operational profits funded most business growth. Bank loans on equipment and property provided additional funding. The company is now an internationally known maker of specialty guns parts and gunsmithing tools. The rural gun enthusiast has retired from the steel mill and moved his home and business close to where he grew up. He is semi-retired from his second business and spends most of his time building boat docks and other projects for fun.

As much as a new company would like to create a revenue stream, it is not always practical to do so from new technology. Frequently, the work needed to go from idea to production is far more expensive than can be funded by the Entrepreneur, family, friends, and personal lines of credit. In such cases, the project will require outside third-party equity financing. Here, an Angel Investors may be the best choice.

Angel Investors are generally individuals or families who have money to invest in businesses or technologies they "like." Some Angel Investors are strictly looking to make money, but most also want to do something "interesting" or "good" with their money. They may take additional risks and demand less return if they really "like" the idea. They may be patient. They might be more helpful and willing to be engaged than the usual investor in making the technology a success. They will be very interested in the Business Plan and will likely be personally involved in assessing the business and its probability of success.

Angel Investors often sound good to the Entrepreneur at first blush. Unfortunately, this can become a nightmare if the Entrepreneur is not prepared. The Angel Investor will want equity. They want to make good money if the idea succeeds. They will also want to reduce their risk. Hence, most Angel Investors will propose rather complex deals with elements of a loan that can be converted into equity. Most Angels want an "exit strategy." An "exit strategy" is a planned future event that allows the investor to get all or some of their money out of the deal. Future events such as bringing in additional investors or "going public" will trigger the "exit strategy." In any case, most Angel Investors will be concerned about getting some or all their money back on some specified timetable. The Entrepreneur must be prepared to discuss these details.

The Angel Investor often seeks partial control in the business and the Technology Development process. This condition can be helpful, but it is usually a mixed blessing – especially if the Angel Investor has no direct experience with the subject industry or developing similar technologies. Even the best of intentions can lead to unintended disaster.

Evaluating the deals proposed by the Angel Investor can be a daunting task. The Entrepreneur should expect the agreement will favor Angel Investor. Here, the term "Angel Investor" can be a serious misnomer. Angel Investors are not above cutting deals that take all value from the Entrepreneur. Therefore, the Entrepreneur must negotiate shrewdly with the Angel Investor, and they are not usually that good at doing this.

> The Entrepreneur should get third parties involved in negotiations of deals with potential investors. Unfortunately, Entrepreneurs often overvalue their technology or business and underestimate the consequences of conversion and exit strategy clauses. The third parties may include an attorney, but attorneys are not good at evaluating business deals. They are experts in the law but not business experts – especially technology businesses.

When dealing with Angel Investors, there is simply no substitute for an excellent Business Plan that includes the Technology Development Program details. If the Entrepreneur has a Business Plan meeting the criteria laid out in this book, they will have the content to communicate with Angel Investors. Initial communication will be simplified summaries of the whole plan. Eventually, the Entrepreneur should share the plan with any serious potential investors. An up-to-date, well-documented Business Plan will protect the Entrepreneur from fraud claims and probably from having the wrong Angel Investor. On the flip side, an Angel Investor looking at a technology-driven, fledgling company that does not see a Business Plan like described here should probably run away from any deal as quickly as possible.

After family and friends and the first serious money from the Angel Investor(s), the monetization plan for the technology may be ready for a significant influx of capital from Venture Capitalists (VC). These are usually groups of investors who have put together a fund managed by a known colleague. Most VCs are looking to increase value rapidly. Their time horizon is generally about two to three years. They generally stay in industries known to them, but their goals are often strictly financial. They want to "triple their money in five years or less." Hence, they are looking for more than 30% ROI with a well-defined exit strategy. In addition, they frequently want to take the company to an Initial Public Offering (IPO = publicly traded stock) to have a ready market for their investment.

High-rolling, VC financing is appropriate only with a fully developed and validated Business Plan. VCs are on a strict schedule, and if progress starts falling behind, the VC will take action to force short-term monetization. The actions taken by the VC will not necessarily be in the best, long-term interests of the company or the technology. They often cause a high-risk expansion, merger, or even force an immediate asset liquidation. They usually take the approach that getting what they can now out of a mediocre deal is better than waiting. They want the cash now so that it can go back into "circulation" and get that 30% + ROI elsewhere.

Bringing in a genuine VC is a considerable risk for the Entrepreneur. Something big will happen in two or three years, and everyone needs to be ready. It is sometimes not easy to identify the VC mindset. They have been the participants in numerous failures and scandals. Hence, they often don't associate with the term "venture capital." They sometimes try to look like an Angel in demeanor. Therefore, the Executive needs to look past the "labels" and evaluate the potential investor for their degree of VC tendencies. The best evidence is how the investor has behaved in the past and what specific "suggestions" they propose.

The VC invests early in many "good-sounding" ideas that are not necessarily financially viable. Most importantly, the scheme must be "fundable." That is to say, the idea must have enough "sizzle" to attract other investors. The VC strategy is to invest in many of these "hot deals," expecting a few to pay outrageous returns. Of course, the others will turn out to be losers, but that's not important. Instead, the plan is to "dress up the pig" with early money and get it positioned to attract more investors. These steps multiply the "value" of the VC's initial investment. The VC can sell off some initial investment to later investors (especially at an IPO) at higher rates, reducing or eliminating their cash at risk. If the business fails, VC losses are negligible, but the rewards can be astronomical if the company hits it big.

The VC is only looking for a few out of 20 deals to hit big. By shrewd financial maneuvering, the VC can minimize or eliminate their losses. But what of the later investors? Will they be as lucky or savvy in getting their money out before the "house of cards" tumbles? Some will, but most won't. Unfortunately, the majority of the 20 deals that fail may represent a small army of knowledge workers who have invested vast amounts of personal time and expertise into what turns out to be a financial "flim-flam."

The key tells of a VC are:

- Always in a hurry,
- Not that interested in the evidence that the technology "works,"
- Focused on the "story," the "sell," and "appearances,"
- Looking for multiple "rounds" of investments to bring in others,
- Loves to use bogus rules-of-thumb for business valuation, and
- Looking for quick ways to sell, trade, or flip their investment.

The Executive must be careful not to engage with a VC until ready to "bet the farm." Delays will turn the VC into a liability, as will the "promises" they made to other investors and stakeholders. The "failure" created by trying to rush forward will be challenging to overcome, and the Executive and the Technology Development Team will probably lose everything.

The VC approach is often not the most effective way to fund a new technology – especially at the earliest stages. Nevertheless, when combined with more patient sources, VC funding can be indispensable. Example 28 shows what it looks like if things work. Jeff Bezos used a combination of personal money, family money, Angels, debt, and VCs to go from concept to an ongoing concern. Note that the simple table in Example 28 does not reflect the many challenges faced by Amazon. Missing are numerous twists and turns such as the "dot.com bubble" of 2000, lawsuits with big players like Barnes & Nobel and Walmart, or competition from Alibaba. The company weathered these storms but didn't turn a profit until 2001.

Example 28: Financial Chronology of Amazon.com (adapted from Winning Angels [14])

Price/Share	Date	Action
$.001	Jul 1994–Nov 1994	J Bezos Invests $10 K and borrows $44 K
$.1717	Feb 1995–Jul 1995	Family Invests $245 K
$.1287–.3333	Aug 1995–Dec 1996	Two Biz Angels Invest $54 K
$.3333	Dec 1995–May 1996	20 Biz Angels Invest $937 K
$.3333	May 1996	Family Invests $20 K
$2.3417	Jun 1996	2 VC Funds Invest $8 m
$18	May 1997	IPO 3 m Shares Raise $49 m
$52.11	Dec 1997–May 1998	Loan and Bond $236 m to Retire $75 m of Debt and Finance Ops

After engaging family and friends and maybe even the first outside Angel Investor, the technology may not yet be ready to make an "all in" bet on commercialization. There may yet be some technical "warts." The Entrepreneur, the Development Team, and the Angel Investor(s) may be facing a severe dilemma. No doubt much has been spent developing the technology. No doubt the anticipated increase in valuation is lagging the timetable. Undoubtedly, revenue is too low to sustain at the current "burn rate." The Entrepreneur and the stakeholders frequently find themselves at a critical decision point. The temptation to bet on a VC is high, but this could be a disaster. Without additional investment, the technology could die from lack of funding.

Forming a strategic partnership with an established company is a strategy that works with some technologies. The partner could be a supplier, a customer, or a competitor interested in applying the technology. Here the IP strategy may turn out to be crucial. For example, suppose the Technology Development plan is sound, and the IP is well protected by patents or trade secrets. In that case, the value of the technology may be in licensing it to other players with existing cash flow. In any case, we again see how important an up-to-date, realistic Business Plan can be. Proper planning, execution, and documentation can build the value of the IP, giving additional flexibility in monetizing its value.

Another possible source of third-party funding is Crowdfunding. The most successful type of Crowdfunding has been product pre-sales. Pre-sales is not very practical for Technology Development but could be helpful in the later stages of commercialization, especially for certain kinds of new consumer product technologies. Equity Crowdfunding should be a potential source of equity financing for Technology Development, especially for technologies with significant social impact. Unfortunately, securities regulations at the state and federal levels have seriously hampered the value of Equity Crowdfunding. In the interest of Consumer Protection, the SEC has made Equity Crowdfunding nearly useless for its intended purposes. We will return to this topic later with some suggestions.

The Institutional Investor is the final source of funding we will present. The Institutional Investor can be a wide variety of investment groups, including investment banks, pension funds, and even university endowments. The Institutional Investor usually facilitates the transition to secondary markets with marketable securities. As a result, they are more likely to fund an Initial Public Offering (IPO) than make a private, non-tradable investment. Again, having ample evidence on the development of the technology will be crucial with these large groups – especially if the next step involves taking the company public in an IPO.

We have talked about some investor types and a general investment timeline. Now we will talk about some specifics of trying to raise capital. We will be talking primarily about third-party sources that will require objective evidence. The reader should realize that it is always better to have more than fewer factual details and evidence, even with family and friends. Squeezing out the unknown and getting to the truth – even when it is not what we want to hear – will always save us heartache in the long run.

Raising capital is a stepwise process that follows a typical pattern. It includes:
- Assessing Capital Needs
- Setting Capital Goals
- Detailing Capital Goals and Strategy
- Identifying Potential Capital Sources
- Structuring Capital Offering(s)
- Managing the Campaign

There are, however, some persistent myths that we should dispel before describing these steps.

Dispelling Some Myths about Raising Capital

There are many myths and bogus information about trying to raise capital. They include:
- It's All about the "Pitch"
- It's All about the Plan
- It's All about the People
- It's All about the Tech
- It's All about the Market
- It's All about Saving Taxes
- It's All about Luck

Some will tell you that raising money is all about having a slick "pitch" that you can rip off in five minutes or less. These are the same folks that hold pitch contests and award prizes for the best pitches. Well, that may win ribbons and even make the news, but a slick pitch doesn't make for a good company, and very few pitch contests result in actual investment. It is way overhyped.

Others will try to convince you that it is all about having a great Business Plan. I tend to get trapped in that thinking, but it is incorrect. Without a realistic Business Plan, it is all "hot air." Nevertheless, a practical Business Plan implies solid content backed by evidence. It is more than just a fine story told convincingly. A real Business Plan summarizes initial research corroborated, impeached, and modified by testing, validating, and criticizing. It outlines what is known, suspected, and yet needs to be learned.

Some business pundits believe that having the right people in place is all that matters. Others think it is all about the "killer app." Still, others will focus on how the technology or the "deal" will save taxes. But we have all seen "dream teams" lose the big game, the "killer app" that no one wanted to buy, and the tax dodge that turned out to be vaporware.

And of course, there are the devout cynics that think everything is just a matter of luck. The truth is that all these factors and many more can become significant. They can all be contributing factors that can enhance or diminish the chances for success – especially when our goal is to create a sustainable business. Predicting long-term success in the Technology Development business is difficult, but succeeding pays vast dividends for owners, employees, investors, and many stakeholders. Furthermore, improving success rates in finding, developing, and commercializing new technologies lowers risks, frees up more capital, and improves returns on the labor invested in Technology Development. Thus, enhancing success in Technology Development grows wealth and enhances the distribution of wealth simultaneously.

Assessing Capital Needs

There is no objective or responsible way to assess how much money will be needed to develop and monetize a technology without a realistic Business Plan. Hence, before seeking capital, the Entrepreneur is morally obligated to create and vet a practical Business Plan that includes the many issues discussed earlier. This Plan becomes the objective basis for estimating the capital required to monetize the technology under development. As we have stated earlier, the quality and reliability of the Business Plan will improve over time.

New information will continuously disabuse the participants in Technology Development of their wilder fantasies. That is the nature of the Technology Development process. Nevertheless, failure to have an up-to-date, honest Business Plan that includes Technology Development details is ample evidence that the owners, managers, funders, and stakeholders participate in a flim-flam. Some may be self-deluded pigeons, while others may be little more than hucksters running a clever scam. Everyone should run away from deals like this, and those in positions with fiduciary responsibilities should seriously consider exposing the flim-flam to mitigate the financial impacts.

An appropriate Business Plan should provide the information needed to:

- Estimate the money required to monetize the technology,
- Estimate the other resources (personnel and equipment) required to monetize the technology,
- Develop a timeline for both the investments needed and the cash flows to be generated,
- Calculate potential Internal Rate of Return, and
- Identify and evaluate "key success factors" and "deal killers."

Set Capital Goals

Armed with the information from a good Business Plan, the Development Team can begin to create a written Capital Campaign Plan. Many of the resources needed are apparent. It will take people, equipment, facilities, marketing, and working capital. These can be turned into costs by doing some research. This research includes estimating appropriate salaries and bonuses, getting budgetary quotes on equipment, and getting expert estimates on facility costs. The Development Team will need to make and document assumptions about making, buying, leasing, or subcontracting essential resources and activities. At this stage, it would be wise to keep cost estimates generous and avoid being optimistic about future costs.

One of the most important and most challenging to assess is time. Developing the technology and getting it to market will take time (See Axiom 4). Writers, readers, and reviewers of the Business Plan should agonize over the timetable. It is never "right" and always critical.

What should emerge at this stage is the first cut at a financial projection of the Technology Development Program. The next step will refine this crude estimate.

Detail Capital Goals and Strategy

The next step in the Capital Campaign will be detailing the Capital Needs. It is one thing to estimate the total cost of a Technology Development Program and an entirely different matter knowing the details and the timing of those costs. Here we will specify our assumptions about making, buying, leasing, sharing, swapping, or subcontracting for the resources we will need. We will also identify the timing of those costs. Finally, we will specify what we expect in revenue and when we will need cash to keep the program going until we reach our time horizon. If practical, we will extend our time horizon far enough to include investor exit points. Hence, we will have to include proposed investor payout events and estimate payout percentages in our thinking.

We will be creating a complex Financial Plan to fund our Technology Development Program. It will be an iterative process as we make assumptions, make projections, and evaluate results. We will be creating complex spreadsheets able to accept multiple modifications. The Entrepreneur or anyone on the Technology Development Team will rarely be qualified to make these projections. Instead, someone skilled at spreadsheets and forecasts and with a finance and accounting background should create, validate, and update these projections as needed.

Some guiding principles throughout this forecasting stage will be:
- The Devil is in the Details
- Although Plan < > Reality – Detailed Planning Needed
- Equity is the MOST Costly Capital
- Debt is "Leverage" – Multiplies both Good and Bad
- "Partners" are Needed – Selection a Life or Death Decision
- Time is Your Best Friend & Worst Enemy
- Complexity = Risk (Keep it as Simple as Possible)
- Cash is King
- You are Probably too Optimistic

This forecasting step is often an exasperating process. The amount of detail can be overwhelming. Developing and modifying a useful spreadsheet without "blowing it up" is a skill few possess. Most don't have the ability, patience, or discipline to do much more than create a befuddled mess that is worse than useless. The Entrepreneur should engage an experienced expert to do this.

For all the effort that will go into this Capital Campaign Plan and spreadsheet, it is critical to remember that the Plan will not be the reality. Users must update the Plan regularly, or it will become a millstone. The need to frequently update the Plan makes it even more imperative that an expert be engaged and retained throughout the process.

The Plan will make assumptions about equity versus debt financing. Where practical, the company should prefer debt over equity financing. Nevertheless, debt to total asset ratios of more than 80% are challenging to sustain. With few exceptions, every company needs some equity, and detailing the nature of that equity position is complex. The Plan should make some basic assumptions and test them with potential investors. Here is another area where an experienced consultant can be handy. They often know investors who are willing to give informal feedback. Too much complexity here is not warranted because most potential investors will bring their ideas/proposals to the table.

Frequently, the Entrepreneur and the Development Team will not have all the resources needed to monetize the technology. Some of these resources will be nonmonetary. An example might be seeking a "strategic" production partner to use the technology and make a product. Finding that partner will take time and money and should be reflected in the Capital Plan.

As we have mentioned before, the element of time is crucial and challenging to predict. Do not allow too much optimism when it comes to milestones. Once put to paper, thinking tends to drift – constantly in the direction to do the most damage. Investors push for shorter timelines while reality seems to run at a snail's pace. Be cautious about getting caught in the squeeze.

Striking a balance between thoroughness and too much complexity is difficult. Plans that are too complicated give the impression of high risk. Too many independent actions must align to work. Projects that appear too simple look incomplete or cavalier. A good approach is to have a master spreadsheet with excruciating detail summarized on a "rollup" sheet available to potential investors.

Entrepreneurs must guard against running out of cash. Many companies fail because financing plans overlook cash obligations. Businesses that cannot meet payroll or debt obligations suddenly collapse. Sadly, the Entrepreneur is always too optimistic about everything. Development Team members are usually somewhat more realistic but often suppress their opinions. Ferreting out the truth is a role best suited for the expert, third-party consultant who can ask the tough questions.

Once assembled, the Entrepreneur, the Development Team, and their consulting expert will do a preliminary evaluation of the Capital Plan. The plan's first iteration should be equity-only, unleveraged, before tax evaluation. This simple approach avoids arguments about acceptable debt levels and various tax considerations. If the Plan does not show an IRR of 10% or more in a "reasonable" time frame, the Plan is probably not workable. Nevertheless, the Team can now revise the Plan to see if it can be made workable. The ability to edit and re-evaluate is an essential benefit of a detailed Capital Plan.

The Team can usually improve the IRR of a Capital Plan by leveraging with debt. Additional improvements can often be made by looking for alternative sources of needed assets that do not require an equity investment (leases, rents, sharing, swaps, etc.). Some very general MINIMUM guidelines for equity only models are:
- IRR must be >10%–12% with no technical risk
- For Angel or VC money – IRR must be >20%
- For plans with technical risk – IRR must be >30%

If IRR still can't meet the minimum criteria above, the Team needs to look deeper into the features of the Capital Plan. Some ideas include:
- Revenue Enhancement?
- Cost Reduction?
- Reduce Time to Payout?
- Sharing Strategies?

- De-Risking Strategies?
- Segmentation Strategies?
  - Phased Approaches to the Capital Raise
  - Find Investors with High-Risk Appetite for Some Reason – Including Social Good

Can the Team change some of the technology features to enhance the revenue potential? Can the Team reduce the cost of development or deployment? Could the technology be applied in a more profitable market? Can the Team reduce how long it takes to monetize the technology and create a positive cash flow? Can the Team reduce the capital needed for development or deployment by sharing resources with a strategic partner? A supplier? A wholesaler?

Can the Team reduce the technical risk by more extensive testing or trials? Would a strategic partner test and endorse the technology? Would a government agency or non-profit organization do a portion of the development and reduce the total cost to your Technology Development Program?

One final possibility is to segment the Technology Development Program. Is it possible to give a higher payout to an early investor when risks are the highest and bring in other investors when the risks are dropping? If you look at the Amazon example (Example 28), you will see that the early family and Angel investors received much higher IRR than later investors. Higher IRR for early investors is a relatively common method of getting "early money."

The Plan developed should be an internally consistent, workable plan. It will be subject to substantial modification as the Technology Development Program progresses. An attractive side benefit of this detailed work is that there is now an objective basis for discussing the **value** of the technology.

The Capital Plan is essentially a cash flow projection based on a specific technology monetization scenario. It is a particular strategy to turn the idea into a time-dependent money flow. The value of the "company" could be estimated using a risk discounted Net Present Value calculation. The Entrepreneur may be sure that every potential investor will do that calculation using their own risk-based "discount rate." The Entrepreneur should do this calculation and consider it the "back of the envelope" value of the company. From this point forward, the Entrepreneur should have a robust and objective opinion of the company's value that they can quote and defend. The Capital Plan and the company value should be the critical content of any "elevator speech" that the Entrepreneur should always have ready.

> An effective 30-second "elevator speech" is a two-sentence description of what the technology does for an identified market, followed by a sound, evidence-based estimate of company value and a percentage of the company open for purchase at that valuation.

The Plan developed above gives sufficient detail to identify the investor potential. It should clearly describe what investments are needed and what returns are possible.

When combined with the Business Plan, any potential investor should have what they need to decide their level of interest. Of course, potential investors will not want the "whole story" at the first introduction. They will want to be involved in a "buying cycle." The Entrepreneur is "selling" to potential investors from this point forward. The first step is "finding qualified buyers."

Identify Potential Capital Sources

As we have noted earlier, not all investors are alike. They don't have the same motivations. They don't have the same constraints. They don't have the same desired outcomes. Most are:

- Conflicted in their motives,
- Looking to turn a quick buck,
- Not as sophisticated as they seem,
- Poorer than they portray,
- Egotistical,
- Not telling the truth, and
- Not who you want funding your business.

We are looking for an exceptional investor – the "right" investor. We are not looking for just any investor – especially very early in the game. That is why most companies, even the biggest, begin their funding with their own money and effort. The next level is usually family, friends, and close associates. Some of what we say here will apply to these early investors, but our focus will be on those third parties that the Entrepreneur and the Development Team do not know personally. We are now "selling." Like all sales processes, we start by narrowing the field from the many bad choices with the characteristics listed above to the very few good options. We are "qualifying" prospects.

So, who are the "good prospects," and where do we find them? Some of the characteristics are:

- They have money,
- They have invested similar amounts, in similar companies, in similar fields, at similar stages of the company development,
- They display genuine interest in and knowledge of your technology/business,
- They are open to discussions about your technology and are keen to hear the details,
- They are open about discussing their previous investment and business successes and failures,
- The investment decision-maker is personally involved in the transaction, and
- They bring something to the table other than just money.

It is surprising how many "investors" are just pretending to have money. It is an ego trip to be "courted" as a potential investor. Many do nothing more than float from one "pitch contest" to the next to get folks desperate for money to "suck up" to them. The best evidence that the potential investor is "real" is that they have invested in similar deals with similar companies.

The right investor is one that "invests" time in evaluating the technology, Business Plan, and structuring the deal. They have more than money at stake. They often work hard to make the deal into a successful business and exercise patience far beyond any legal or fiduciary requirement. We are looking for personal engagement where "failure" is "bad" and "success" is part of the "reward." The Entrepreneur should listen carefully to what the potential investor says about previous successes and failures. Who do they blame? What do they brag about? Listen, and they will tell you if they are a builder or a taker, a partner or a thief.

> At some point, the technology and the business will "graduate" to where investments are impersonal, financial transactions driven by a vast system designed around secondary markets. The great challenge is getting from concept to IPO. The crucial steps are the first few where third parties become involved. After that, the financial markets start working effectively. They don't work very effectively before the IPO.

Finding the right investor is hard, time-consuming work. First, it requires looking at hundreds of investment entities (especially family offices, Angels, and VCs) whose investment profiles align with the Business Plan. Then, we must cull down those hundreds to a double handful of prospects that need to be contacted and seriously evaluated. This work is a job for the Entrepreneur or a trusted representative. That representative can be most effective if they are a part-owner in the company. Someone who is part of the company is far more believable and less constrained than a consultant or a broker. Consultants and brokers can be very helpful, but they can NEVER replace the knowledgeable Entrepreneur who understands their business, the processes required for effective capital campaigns AND has the wisdom and the discipline to say "no" to the wrong investor or investment deal.

Example 29: The Power of Saying "No"

A startup in the outdoor recreation business was hoping to raise several hundred thousand dollars to acquire their first, company-owned camping facility. They found an investor willing to make a convertible note investment. The deal offered wasn't terrible, but it didn't fund all that the Entrepreneur felt was needed. It just didn't seem that the potential investor would be the best long-term partner. The Entrepreneur somewhat reluctantly passed on the offer. A few weeks later, the Entrepreneur found another investor far more interested in the idea. The new investor could fund multiple sites and was interested in a long-term, strategic partnership.

Finding the right investor is a specialized selling process. Some have tried to "standardize" the approach but with little success. Many charlatans and fools write books, do training sessions, and provide various contests, gimmicks, and simplified methods. They are worse than a waste of time and money – they give investors and Entrepreneurs lousy information. When it comes to rather complex Technology Development Programs, standardized, impersonal processes don't succeed. To be most effective, raising money must follow a disciplined, personal selling process that engages the Entrepreneur, the Development Team, and the investor(s).

There are multiple ways to find well-researched prospect lists that can serve as a starting point. Nevertheless, it takes time and effort to contact prospects with an attention-getting lead and quickly qualify the prospect. Moving the prospect down the selling cycle while being constrained by SEC regulations is an art that few have mastered. It is a complex "consultive sell" with little more than a few guiding principles to assist the neophyte:

Guiding Principles for Contacting Potential Investors

- Finding Investors Is Sales & Marketing
- Finding Investors Is Communicating & Convincing
- Finding Investors Can Have SEC Regulation Consequence (Especially "Advertising")
- Fraudulent Conduct Can Result in Civil and Criminal Penalties
- The Truth Is the Best Defense
- It is Legal to Lose Money – Even to Be Wrong – But Deception for Gain Is a Crime = Fraud

The ideal approach is to have a trained person begin the selling process. This person should be associated with the business but probably NOT the Entrepreneur. The idea is for the salesperson to "close" for a high-level meeting between the potential investor, the Entrepreneur, and members of the Development Team. Therefore, setting the stage for the conference is crucial. Some of the key features are:

- The investor is pre-qualified, (has money, has an interest, can act)
- Any features of a potential "deal" are known (i.e., no giant surprises),
- The "right" attendees will be at the meeting,
- The meeting has a planned selling outcome known by the Entrepreneur and members of the Development Team,
- The Entrepreneur and members of the Development Team are trained in Team Selling and are clear on their roles, and
- The Entrepreneur and members of the Development Team are clear on "walk-away" points.

Structure Capital Offering(s)

A Capital Offering is an offer of ownership in the company to a potential investor in exchange for something of value. That "something of value" is often money, but it could be other assets such as:
- "Free rent" in a building,
- Labor hours,
- Use of equipment,
- Inventory,
- Ownership or use of Intellectual Property (IP),
- and many more.

The "something of value" could have a fixed dollar value, or the dollar value could depend on future events. The "offer of ownership" could be a stock or a percentage of an LLC. It could be immediate ownership or a future ownership option (e.g., a convertible note). The variety of possibilities is almost endless and can cause serious misunderstandings and legal entanglements if not appropriately handled.

In general, the Capital Offering is a formal, written document, but verbal and written communications could be an "offer." Hence, Executives must make final offers in writing that clearly state the written documents supersede all other communications. Executives must be cautious with Capital Offerings. The danger is that a Capital Offering may be creating a "security" – an ownership right or tradable asset or derivative regulated by the Securities and Exchange Commission (SEC). SEC regulations are complicated, and no final Capital Offering should be without the help of an attorney who is experienced in SEC regulations and making small business Capital Offerings. Nevertheless, the Entrepreneur must remain in charge of the process. Hence, we will give a concise primer on SEC regulations that impact the Capital Offer.

The government created the SEC because of the Great Depression. The idea was to protect investors from being cheated by stock scams. The stock market crashed when investors (including the public) borrowed money to buy stocks ("buying on margin"), many of which turned out to be worthless. Companies made outrageous promises to get investors' money. Many of those promises never had a chance of success. A stock-buying frenzy ran stock prices up and appeared to give investors huge profits. When the economy cooled, and it became clear that many companies were not making profits, the downturn created a panic. When banks called the margin notes, a selling spree ensued. The plunge in stock valuation wiped out big investors, small investors, and many investment banks.

Some of the more pertinent SEC regulatory events were:
- Securities Act of 1933
  - The overarching goal is to enable investors to make informed investment decision by requiring full disclosure of all important information concerning

the issuer and its business and prohibiting deceit, misrepresentations, and other fraud in the sale of securities.
  - Requires sales of securities, unless an exemption applies, to be registered with the SEC.
- Securities Exchange Act of 1934
  - This Act created the SEC and provided it with the power to register, regulate, discipline, and oversee firms, agents, and agencies as well as FINRA, NASDAQ, NYSE, etc.
  - Requires public companies to create/disclose annual and quarterly reports and regulates secondary market transactions, insider trading, and other types of fraudulent actions.
- Investment Company Act of 1940
  - Regulates mutual funds and companies that invest in securities for others and offer its own securities to the investing public.
- Investment Advisers Act of 1940
  - Regulates persons who provide investment advice to others.
- Sarbanes-Oxley Act of 2002
  - Created the PCAOB to regulate accounting practices and requires CFOs and CEOs of public companies to swear to the truthfulness of annual and quarterly reports (i.e., makes them personally liable).
  - Mandates that outside auditors use GAAP principles when reporting on a company's financial condition.
- Dodd-Frank Wall Street Reform and Consumer Protection Act of 2010
  - Created the CFPB and reshaped the regulatory oversight of financial institutions as well as provided protections to whistleblowers.
- Jumpstart Our Business Act of 2012
  - Created to streamline certain capital raising avenues by minimizing regulatory requirements.

It is also important to note that each state has its own regulations. These generally coordinate with the SEC and monitor smaller transactions. In some cases, more flexibility is allowed with smaller, intrastate offerings.

The primary goals of SEC regulations have been:
1. Provide investors with reliable data to make informed decisions,
2. Limit the investor pool to those who can make informed decisions, and
3. Require companies and their representatives to report relevant financial performance reliably and accurately to investors.

In most cases, the Entrepreneur will want to avoid Capital Offerings that require registration with the SEC. The paperwork and reporting burdens are often too cumbersome for the small enterprise. The Entrepreneur may also want to avoid launching businesses in states with burdensome regulations on small offerings exempt

from SEC regulations. Nevertheless, we propose that Entrepreneurs meet the basic disclosure guidelines of SEC regulations for the following reasons:
- Informed investors are better investors – they are more engaged and apt to support new opportunities that emerge and excise patience when challenges arise
- Investing time and money in poor ideas is a waste of valuable resources for everyone
- The disciplines required to meet SEC goals are some of the disciplines necessary to have a good company
- Meeting the basic SEC goals from the start will make a transition to a public offering (IPO) much more straightforward and likely of success
- It is the right thing to do

Hence, we recommend that the Entrepreneur treat raising money as SEC regulated. In other words, the Executive should:
- Disclose risks,
- Report financial data honestly,
- Work with sophisticated or accredited investors,
- Avoid public advertising of offerings,
- Create investor relationships, and
- Write formal Capital Offerings that spell out expectations and disclose risks much like a formal Prospectus would do.

Above all, the Executive should keep things as simple as possible. Cutting corners, leaving things "fuzzy," and creating complicated relationships will not go well.

Example 30: Complications = Risks

A startup chemical processing company funded its ongoing research by selling licenses to use a patented, partially developed process. These actions created a complex portfolio of license arrangements that were difficult to maintain. When a strategic investor came along who wanted to fund the commercialization of the patented process, untangling the multiple relationships became a chore. It resulted in legal disputes that consumed time and money, inhibited commercialization, and soured internal relationships.

There are just a few guidelines for structuring a deal. There is no single answer, no single formula. Investors will often begin their thinking with the value of the company. At some point, they will look at how much money the company can make and how much money the investor can get out of the company. They will compare that to what they think they might have to pay to get that future benefit. Finally, they will factor in perceived risk and decide on a money value. The Entrepreneur needs to think like an investor to "sell" the deal to the investor. The Capital Offering should appeal to the typical investor. It is, after all, an "offer to sell."

The Capital Offering will, in most cases, just be a starting point for a "negotiated sell." Hence, putting together a detailed Capital Offering as a final offer is unrealistic and counterproductive. Although we want to think and speak formally at the right time, those who act too formally early in the process will never raise money – at least not from the kind of investors the Entrepreneur needs. What is required is the engaged, informed, patient investor. Nothing else will work well for Technology Development. Unfortunately, these kinds of investors are hard to find, and Entrepreneurs often don't want investors engaged in "their business."

The tension over "control" is a dilemma for many Entrepreneurs and a source of frustration for their stakeholders. Entrepreneurs, especially those developing technology, NEED investors who will engage. Technology Development is a long process. The path is difficult to predict and will always take unexpected turns. Hence, the Entrepreneur needs patient and engaged investors. Yet, the Entrepreneur is often the loner who marches to their own drumbeat. Although this is frequently a virtue, finding investors can be a serious flaw in the case of finding investors. The Entrepreneur who is developing technology must temper their independence when finding an investment partner. They may need the help of a trusted advisor to do this.

The Entrepreneur must come to grips with:

- What they want out of the investor deal, and
- What they can "live with" out of the investor deal.

Many Entrepreneurs want too much control and freedom for most investors – and often for their own good. They also value their opinions too highly and frequently poison negotiations and relationships with investors in a clash of egos. The Entrepreneur will have to strike a type of "Faustian Bargain" to get the money and time they need to develop their technology into a commercial success. That starts with having a reasonable "ask" as a generic Capital Offering. If successful, the process will culminate with a relationship with an engaged investor who understands the challenges and is committed to seeing the project through to financial success. Frequently the engaged investor will supply the required discipline to transform a good idea into a great business. Entrepreneurs need to learn this.

Manage the Capital Campaign

One of the things we tell our clients is that raising capital for Technology Development is an iterative process. The challenges and opportunities are constantly changing as new information becomes available. Therefore, managing the Campaign should be viewed as a Continuous Improvement Process. There will be times to go out for additional funding and times to circle the wagons and rethink. The smart decision could be to halt a Capital Campaign while new information is being collected and processed. Hence, active management of the Campaign is an essential ingredient.

The Entrepreneur, the Development Team, essential consultants, and other stakeholders (including existing investors) should meet regularly to update the Business Plan, the Technology Development Plan, and any ongoing Capital Campaigns. Unfortunately, many companies are reluctant to hold these kinds of meetings. It often seems that many are afraid that "bad news" might be shared that could derail the project – perhaps even destroy the company. Although possible, this rarely happens. More often, the "bad news" not shared becomes the missed opportunity to succeed.

Example 31: The "Bad News" That Could Have "Saved the Day"

In an earlier example, we saw a high-pressure steam injector was too energetic to work well in a corn ethanol application. It broke too many chemical bonds too quickly. That overwhelmed the yeast. This "bad news" essentially doomed the small business attempting to commercialize the technology in the corn ethanol business. It turned out the R&D group had run across two other applications that looked very promising. The group had learned that the injected steam produced high-energy free radicals through steam cavitation. These radicals broke down toxic chemicals and pathogens. Had the company pursued one of these two applications, the company might have been a commercial success. Unfortunately, the opportunities were buried in R&D while the company "bet the farm" on an application that failed.

Chapter 7
The Future of Technology Development

The twenty-first century will require many innovations – especially new, disruptive technologies. The world population grows at unprecedented rates while concerns about supporting the growing population also grow. When added to emerging concerns about environmental damage and depleting resources, the situation is alarming to many. Although there is a great need for new, sustaining technologies, there is no consensus. Current collective thinking is that our new technologies will arise from public/private, cross-cultural collaboration. That has never been the case throughout human history. Unfortunately, most new technologies have been born out of conflict, often of the most violent and inhumane sort. Long periods of peace have frequently ended with technology stagnation, leading to many ills, including tyranny, overpopulation, disease, and expanding poverty. These conflicts have led to a realignment of "winners" and "losers." As disturbing as these conflicts and upheavals are, we humans still seem subject to Malthusian factors that are a part of "normal progress." Hence, we should not expect our future to be somehow miraculously free of conflict. However, we would be wise to limit the conflict to "wars" of economics, innovations, words, and ideas rather than weapons of mass destruction.

Earlier, we talked about how Technology Development can contribute to economic wealth and wealth distribution under the right circumstances. We will now look at some of the specific critical Technology Development challenges facing our culture today. Several of these are chronic problems that have been around for centuries. The irony is that the first "opportunity" we will discuss is a side effect of one of our great disruptive Technology Developments – the oil and gas industry.

Critical Technology Development Needs

Among our critical problems that will require disruptive technology solutions are:

- Increasing carbon dioxide (CO_2) levels from the burning of fossil fuels
- Pandemics and other healthcare challenges
- Explosive growth in population – especially in poorer and developing nations
- Weapons of mass destruction in the hands of lunatics

Many subsidiary issues pale in comparison to these. Non-profit activists inundate us with requests for our money to stamp out legions of ills. Among these are poverty, homelessness, malnutrition, mental illness, pet abuse, obesity, racism, illiteracy, and a thousand other worthy (and some not so worthy) causes that tug at the heartstrings. These programs make suitable tear-jerker ads and "feel good" activism, but they have little to do with threats to humanity. In many ways, they suck up resources

https://doi.org/10.1515/9783110451634-007

chasing after things that cannot be fixed at all or have reached the point of diminishing returns. Much of the "investments" made in these are over-the-top wasting of precious capital.

Perhaps our top priority should be the increasing concentrations of CO_2 and energy-related greenhouse gases in the air. The rise in levels of CO_2 from the burning of fossil fuels is undeniable. I have seen levels increase from around 330 ppm to approximately 440 ppm in my lifetime. Early in my analytical chemistry career, I used the CO_2 in the air as an approximate standard until it became evident that it was not steady and increasing year after year. The effect that CO_2 levels have on average global temperatures is more challenging to show because there is much variability in the data from around the globe. Although the evidence in the short run is weak and models developed from the data disagree, historical data strongly support the effect, as do laboratory experiments that date back about 150 years. When taken together, the data indicate that CO_2 levels are rising and that average temperature will continue to rise as a result.

Unfortunately, that is where the science ends, and the propaganda begins. The dire predictions that we would die in 12 years were complete nonsense. It was the propaganda of self-serving pseudo-scientists and rabid econazis hoping to stampede the public into supporting their peculiar vision of eco-dystopia. The earth functioned quite well with much higher CO_2 levels and global temperatures than we now have. Many interior land areas were much better off in the warmer, wetter, greener climates of the past when CO_2 levels were much higher than today.

However, dismissing the vast disruptions of a warmer, wetter climate is not reasonable. Sea levels will rise, and storms will be more frequent and violent. These changes will displace humans from the coasts and low-lying areas. We will have to adjust to these new conditions. Unfortunately, the technologies we are chasing today are, for the most part, naïve, counterproductive schemes that will enrich a few (especially academics) and do almost nothing for the rest of us. They are a dizzying array of energy-intensive solutions that will probably make CO_2 emissions worse in the short run. They have little impact on the current climate trajectory, will impoverish the societies that are silly enough to implement them, and misappropriate the capital needed to deal with the real problems that are on the way. It is one giant "eco-scam."

What we need to address the challenge of CO_2 and other greenhouse gas emissions are disruptive technologies that:

- Have high energy density and low, long-term cost – such as atomic power,
- Improve energy transmission efficiency – especially electrical energy,
- "Harden" infrastructure so it will stand up to more violent weather, and
- Significantly reduce the costs of housing so that citizens can readily relocate.

The USA is making almost no progress in these areas. Instead, we pour resources into existing technologies and hair-brained schemes often spawned by advocacy groups, lobbyists, and universities. As a result, we see an emerging "money-grab" typical of how governments and political donors respond to challenges.

Moving away from fossil fuels and their long-term environmental consequences will require vast amounts of new energy to fill the void left by ceasing fossil fuel production and accomplish the transition to something else. Solar, wind, biomass, and hydroelectric are expensive, have a low energy density, and are unreliable. They are unlikely to provide the energy needs for a growing world. The only energy source on the horizon with the energy density needed is nuclear power. We will need many technological breakthroughs to make nuclear a safe, reliable, and low-cost energy source. It is unlikely that the processes used in the nuclear weapons industry (focused on enriching uranium and synthesizing plutonium) will ever be sufficiently clean, safe, secure, and flexible to be a long-term energy solution. It is more likely that a thorium breeder reactor will be a better solution. There is much more thorium than uranium, especially when considering the isotopes that will support fission. Furthermore, the thorium breeder cycle produces less waste and almost no threat of putting nuclear weapons materials into the hands of terrorists. There is still much work to do, but small (even locomotive size) thorium breeder reactors appear possible.

Almost every scenario proposed to reduce fossil fuel consumption in transportation and manufacturing includes distributing vast amounts of electrical power over large areas (e.g., charging stations for electric vehicles). The amount of electrical energy that needs to be transmitted will increase by a factor of approximately five or more if we are to replace most fossil fuel use with clean electric power. The distribution systems needed to transmit this power must be much more extensive and "smarter" than our current systems.

Since many sources and loads will be variable and direct current (DC) rather than alternating current (AC), we will need extensive new facilities. Some will be DC storage facilities to balance loads, while others will convert DC to AC and vice versa. Any DC storage facilities will need to be close to the points of use because the transmission of DC over distance is very inefficient. One of the significant breakthroughs of the early twentieth century was the commercialization by Westinghouse and General Electric of Nikola Tesla's concept of using high voltage AC for efficient transmission. We will need to develop rapid, high-power, high-efficiency conversion and switching equipment and new storage technologies. The electric grid of the near future will look much different than the current system of a small number of large gas and coal-fired power plants transmitting AC electricity hundreds of miles at 345,000 volts. Predicting what that will look like is impossible – we don't have those "winning" technologies.

We may conceive of potential solutions for the new electric grid by looking at a variety of bench and pilot scale solutions, but only "hardened" solutions will turn out to be practical. Our existing, highly reliable grid is already struggling to meet demands caused by increasingly violent weather and wider imbalances between power, voltage, and phase requirements of sources and uses. The grid of the future will require **greater** reliability. Furthermore, the sophisticated detection and control systems must be "hardened" against physical and electronic attacks by enemies and anarchists. Again, the future grid will be much different from today's grid. The number of "inventions" needed and the testing required to ensure the high reliability of a future grid will be a considerable challenge.

It seems unlikely to balance future energy needs and production without breakthroughs in energy efficiency and conservation. The variety of potential Technology Development programs in this area is staggering. It includes improvements in manufacturing, distribution, and use of housing materials, building and roadway materials, manufacturing processes, energy management systems, shipping processes, and the list could go on and on. The technical challenges are daunting, but the commercial opportunities are encouraging.

Perhaps the second most crucial threat to our future is the danger of pandemics and other healthcare-related issues. *Homo sapiens* are under attack by a dizzying array of viruses, bacteria, and parasitic organisms. Squalid living conditions in many areas of the world (even developed countries), relatively cheap methods of rapid travel, bohemian lifestyles – especially in the West, and cavalier attitudes toward disease among researchers and the general populace, present an explosive environment for the development, mutation, and communication of a wide variety of pathogens of both natural and synthetic origins. Our recent experiences with several strains of the flu, HIV/AIDS, Ebola, and SARS1 and 2 (i.e., COVID-19) are concerning.

Many breakthroughs are needed to reduce the probability of severe outbreaks and to mitigate the impacts when they do come. Of course, many of these will be in the medical and pharmaceutical fields. The development of the COVID-19 vaccines was an impressive feat. We will need to repeat breakthroughs like that at an ever-increasing pace. Hence, we need to develop new, non-invasive ways to detect pathogens, monitor their spread, and design more effective countermeasures. As we deliberately cram more and more people into smaller and dirtier spaces, we will see a broad range of new pathogens and will have less and less time to react before millions are affected.

Medical technology development will not be enough. We need to stop stacking people on top of each other like so many microbes in a petri dish. Instead, we need to spread people out to reduce the speed at which pathogens spread and mutate. We also need to buy time for the medical community and political entities to react to the next pandemic or healthcare crisis. That will require disruptive infrastructure breakthroughs to reduce the cost of housing, water purification, waste disposal and recycling, energy, and transportation of people and resources. Almost everything we are promoting today – from the "Green New Deal" to "affordable housing" to the latest "reality" TV show – has the effect of INCREASING costs, increasing population densities, and encouraging "risky" lifestyle choices.

We should favor technologies that result in distributed living conditions and eschew those that encourage increasing population densities. We should be working on cheap, reliable transportation systems that enable moving away from the cities.

We should look at every crowded bus, train, airplane, mall, and apartment complex as a potential ground-zero for the next mass extermination event.

Closely related to preparing for pandemics is responding to the chronic issue of overpopulation. The primary responses to overpopulation will be political, social, and economic. Unfortunately, silly and self-destructive policies in this area abound. Dealing with them is beyond the scope of this book. Nevertheless, the lack of progress in dealing with overpopulation – especially in some of the neediest countries – will continue to cause increasingly dire stresses on resources and resource distribution. We will need to become much more efficient at turning natural resources into consumables. We will need breakthrough improvements in mining, agriculture, manufacturing, transportation, energy, and education. None of this will happen unless our educational system provides high-quality, motivated innovators and knowledge workers.

Education, of course, has long been a challenge and differentiator between successful and unsuccessful societies. As the world becomes more tightly connected, it seems that allowing abject poverty and ignorance in some sectors of the world is becoming more dangerous. In short, ignorant people are dangerous people. They are frequently impoverished and desperate. They are also gullible and will follow crazy schemes. There is a desperate need to make good education widely available at a much lower cost. Improvements in educational Technology Development will be a significant part of the solution.

Today it seems that an outbreak of Ebola or Marburg in central Africa or an escaped virus from a lab in urban China could rapidly infect much of the world's population in weeks, if not days. Furthermore, the speed by which pathogens travel between population centers increases the chances that the new pathogen will have high mortality. Virulent pathogens tend to burn out or mutate to less lethal and more infectious strains over time. In general, rapid transit combined with dense populations facilitate the spread of more virulent pathogens and contribute to overwhelming medical, economic, and political systems. Hence, improved healthcare, sanitation, housing densities, and public awareness of health issues will become increasingly important. Delivering practical and accurate health information to remote and backward areas of the world is becoming a high priority.

We will not improve worldwide healthcare by fiat, propaganda, or using the current, outdated education methods. To make healthcare information "believable" and likely to be instituted requires more than just a simple PR campaign. People groups need to have background information in science, technology, economics, and politics before voluntarily implementing the healthcare, sanitation, abstinence, and isolation methods required to prevent and effectively respond to new pathogenic challenges. Hence, the challenge is much more than just making public emergency announcements. The public needs to have the background information to understand the data, accurately judge the report's veracity, and devise practical, effective, and culturally appropriate measures. In general, everyone needs to become MUCH more educated in critical topics like microbiology, sanitation, and healthcare.

> Note: We cannot limit our discussion to those "other third-world countries." The response of America to COVID-19 was a shocking display of ignorance.

But even that is not the end to the educational challenge. We should consider many other topics that affect health and general well-being. We need to disseminate additional academic subjects ranging from general literacy to labor skills, from agricultural technologies to linear algebra. Our current method of person-to-person classroom teaching is not getting the job done and probably never did. It is slow, expensive, and complicated by multiple political and labor relation issues. With these old-fashioned methods, the quality of the education is difficult to monitor and more difficult to control. The basic approach to education (especially higher education) hasn't changed since the Middle Ages. Furthermore, our recent experience with COVID-19 has taught us that contemporary Internet-based teaching methods are probably worse than in-class teaching in many, if not most, subjects.

We certainly need additional access to high-speed Internet connections and high-speed personal computers. But, we need much more than that. We need breakthrough redesigns of teaching methods that use the advantages of computer-aided teaching and minimize the weaknesses. I have taught both in-class and online courses. For the most part, online courses are poor. They were patterned after the classroom experience but cannot duplicate that experience. Good online learning must have a thoughtful design that emphasizes frequent understanding testing.

Furthermore, online curricula need to be tested objectively – even harshly – for effectiveness. Unfortunately, even though testing protocols are well known and relatively easy to perform in an online environment, we rarely subject teaching methods to objective scrutiny. Historical, political, and labor relation interests frequently block objective review and performance improvement.

Educators designing online curricula must understand the subject matter, especially how essential concepts build on each other to create a well-grounded basis for student understanding. Furthermore, educators need to continuously update and improve their methods based on objective evidence. Unfortunately, most educators do not have that level of subject matter understanding nor the online training needed to create such curricula. We can attribute some of the failures we saw with COVID-19 to ad hoc approaches thrown together in a panic, but that doesn't tell the whole story. Higher education had been using online learning for many years before COVID hit. As a result, we should have been better prepared.

And finally, online curricula must become much broader in their scope. Online curricula can and should go far beyond just math, science, and language. It should also teach practical life and employment skills. The public needs to know how things get done in modern society since they prioritize improving their quality of life. The graduate should understand and appreciate the contributions of the mathematician, plumber, aeronautical engineer, welder, politician, doctor, philosopher, poet, educator, parent, and artist. Their contributors to culture should be our heroes rather than the thugs, pimps, and tramps often lionized in pop culture.

If we are going to improve education in remote and underdeveloped areas, we need to develop disruptive educational technologies. All curricula will require redesign and coordination around the strengths of online learning. Those curricula should include "hands-on" training and visualization. Virtual reality can do much of this, but coordinating with local or itinerant subject matter experts can enhance the online learning experience. Furthermore, we will need to change how we train teachers, accredit programs, award degrees, manage school boards, and fund education. Education in the future must look much different from what it is today in almost every aspect.

And, of course, all these remarkable new technologies will be for naught if some lunatic blows up the world. I say this a bit "tongue in cheek" because it is not very likely that any one person can destroy the world. But, a small group of folks, a small minority of the world's population, has made life on earth miserable in the past and could do so again. So here we have something of a dilemma. Technology development can improve life, but much of it has destroyed life. So we will need to continue developing technologies that help us both find the lunatics and defend ourselves from the worst they can do.

Surveillance and security technology and the engines of war have been a part of life since the throwing of the first rock. Although there is no evidence to support COVID-19 was part of a military plan, its rapid spread and devastation illustrate what could happen with a viral agent. Rapid detection of viral and bacterial agents will be very challenging. Tiny amounts of a pathogen (sometimes as low as ten virus particles) can pass an infection. Current methods often require replicating ("growing") the virus or bacteria to masses that can be identified and quantitated in a laboratory. This process can take from minutes to weeks to perform. Testing infected persons are more straightforward because they often have tens of thousands of individual pathogens in a single sample. Accompanying symptoms may also help with the initial identification of the specific pathogen. Unfortunately, these agents can be challenging to detect before and shortly after release. Infectious agents can, of course, spread as infected persons "grow" more agents and carry it elsewhere.

Chemical agents are easier to detect than biological agents since most toxic agents require higher concentrations to be lethal. Minimum doses run from a few milligrams (ricin, fentanyl) to dozens of milligrams (VX, GB). Furthermore, unlike biological weapons, chemical agents are not "grown" and passed by infected hosts. Hence, the entire mass of the toxic agent must be "delivered" by some means. Many pounds or even tons of the agent must be delivered over a target area for the agent to be an effective weapon of mass destruction. These methods of delivery may be detected and interdicted. Radioactive and explosive threats have slightly higher mass requirements than chemical toxins and unique delivery systems. The variety and variability of the threats will require a wide range of potential countermeasures. We must continue to develop more effective and reliable methods to sense danger and respond to them.

Our modern infrastructures' high systems integration and automation have made them vulnerable to direct, targeted attacks. These include electronic attacks such as computer viruses and malware. They could also include using the physical and biological hazards listed above strategically aimed at infrastructure. Dealing with the possible threats will be an ongoing challenge that we will meet with new technologies, many unknown today. Key factors will be balancing the cost of developing and maintaining these new technologies with their ultimate utility. We will naturally want to reduce labor costs and human risks in deploying and maintaining these technologies. In general, we will turn more and more to robotics and automation in both surveillance and countermeasures. At the same time, ensuring the reliability of these systems will become an increasingly difficult challenge – especially for those systems difficult or impossible to test. We have gone to war over bogus intelligence at least twice in our history (Spain and Iraq). We should hope for better results in the future.

When it comes to "spy versus spy versus spy," disruptive technology tips the scales in favor of one group or another. We often spend too much money on existing, high-cost systems only to get marginal improvements. A good example would be spending on battleships between WWI and WWII. On the other hand, decisive technologies often languish for lack of funding. These include the airplane in WWI and radar in WWII. Since surveillance, security, and war technologies are competitive at the highest level, the essential breakthroughs are usually among the least expected.

Capital Sourcing Challenges

Many of the technologies listed above have been either underfunded or ineffectively funded. Some of it has to do with short-sightedness. Some of it has to do with a lack of consensus in priorities. Some of it has to do with the unpredictable nature of a "winning" technology. Nevertheless, it seems to me that we have structural issues that make it challenging to utilize capital effectively in these critical areas. These disruptive technologies will find funding difficult with some of the most common sources: government, academia, and big business. These groups often focus on the *status quo* and are strangely uninterested in the critical technologies of the future. I see four major categories where we have capital sourcing issues:

- Evaluation of technology – especially the competency of government programs,
- Use of capital – especially setting priorities in government programs and private investors exercising appropriate patience,
- The scarce supply of capital – especially for primary investments,

- Accountability for government program decision-makers as well as investors and business owners, and
- The patience of capital investments in disruptive technology.

Of course, government policy and funding strongly influence directions in capital markets. For example, government grants, loan programs, and tax breaks attract capital and influence decision-making, even in the private sector. Hence, we must influence government behavior and public opinion to influence capitalization. We will address policy issues in Chapter 8 of this book.

Evaluation of Technology

As discussed earlier, government programs can be an essential source of funding for disruptive technologies but are inconsistent in setting priorities and administering their programs. Government programs reflect the political narratives of the last election cycle. All candidates must declare "for" or "against" various hot issues. They must defend these positions in public debate and promote them once in office. Often the political narratives reflect the donor base – the "beautiful people," big business, labor unions, the financial sector, their local constituency, their party's demographic base, an influential academic institution, or perhaps even key foreign donors. All of these can apply dysfunctional influence on government funding processes by "lobbying" for "pet projects" rather than following a consistent strategy aimed at developing critical technologies.

Government programs often reject some of the most viable technological solutions and too often fund embarrassingly naïve "non-starters." Career bureaucrats who are frequently out-of-date and unaware of the most promising possibilities usually manage the most crucial programs. Unfortunately, special interests within the government department with "vested interest" in other technologies often influence key decision-makers. Some of the worst offenders are scientists in national laboratories working on older, outmoded technologies. Poor government management is especially problematic when government funding activities have weak or incompetent third-party evaluation and review teams to assist in objective technology selection. Too often, the result is the funding of a "hot deal" that turns out to be "hot air."

Especially concerning is the new relationship between the government and academia. Since WW2, new collusions similar to President Eisenhower's military/industrial complex developed between the government and academia. Government funding of higher education began with the GI Bill (1944–1956) and continued to grow with several loan programs beginning with Sallie Mae in 1958. We now see many government programs that include contracting set-asides to universities. Some government programs forced those hoping to develop new technology to "team" with a university, giving the university a significant role in the process.

In my experience, the academic community in the USA has not been very effective in Technology Development outside of medicine and, to a lesser degree, computer applications. With a few exceptions, universities are disconnected from markets and not qualified to develop technologies. One exception may be the field of medicine. Many teaching hospitals treat patients directly. Hence, they have a connection with their "customers" and are more likely to focus on valuable developments.

Example 32: The Worst Patent Application Ever!
I am an inventor on six awarded patents. I have worked on about 20 different patent applications. The best patent attorney I ever saw had a Ph.D. in chemical engineering along with his JD. By far, the worst example I ever saw was a senior patent attorney for the technology transfer group of a major university. The patent development team gave him an excellent summary of the process. The first draft that came back was a 43-page hodge-podge laced with grammar, spelling, and logic errors barely resembling the process. Thinking that this must have been some strange clerical SNAFU, we met with the attorney on a conference call. To our horror, we discovered that the crazy written mess was a good sampling of how his mind worked and what the final product application would look like if left up to him. Unfortunately, the technology transfer contract required that we go through the university for patents on technology that we developed related to the original technology transfer project. We had to struggle through the process with this attorney. The development group finally gave up on that patent.

Another concern is the cozy relationship between the government and non-profits. In the late 1960s, two social scientists, Pearl and Riessman, in their 1966 book, *New Careers for the Poor* [21], claimed that government/non-profit partnerships could reduce urban poverty. Pearl and Riessman contended that non-professional service jobs created by government-funded, non-profit organizations could employ the inner-city poor. In a 2019 article, Claire Dunning of the Washington Post challenged the results of that experiment claiming that the low wages of non-profit workers have exacerbated poverty in many urban areas [22]. We now see many government contracts (especially local economic development contracts) requiring the participation of non-profit organizations, often as the lead. Non-profits now compete with private industry for government contracting dollars and control of Technology Development. Interestingly, many grants and gifts to non-profit organizations are in the range that small, Technology Development businesses need to go from family and friends financing to a sustainable business – $250 K to $5 million.

> Note: It is also possible that the construction of urban non-profit facilities – especially those attracting professionals – can contribute to "gentrification" and homelessness in some communities.

The determining factor in the successful technology evaluation may be closely related to moral hazard. Government bureaucrats, academics, and non-profits are not likely to be held accountable for the resources they blunder away on "non-starters." They can repeat mistakes many times before facing the consequences if they follow strict contracting rules. With good spin skills, some bureaucrats can turn a major disaster into a promotion. Academics can often point out what has been "learned" and chalk up big price tags to the cost of "education." Non-profits are far more about valiant efforts toward lofty goals than creating sustainable sources of wealth. Indeed, terrible blunders can happen with private funding, but significant financial errors are rarely ignored in investment circles and could result in legal actions with serious – even criminal – consequences.

Private investors are often better at evaluating technologies and participating in Technology Development than are government agencies, universities, or non-profits. They are certainly more "at risk." They are, however, often limited in their expertise. As we have mentioned before, it is essential to find those investors with adequate familiarity with the target industry or market. Where that familiarity exists, the private fund will generally recognize a good idea and will probably have a realistic view of how long it will take to commercialize it. Where that expertise does not exist, blunders are likely.

And finally, though big business has the expertise and the financial ability to fund new technologies, they are often not interested in critical disruptive technologies – especially at early stages. Big business is often fully invested in developing incremental improvements to their existing technologies supporting their core business. As a result, they have little appetite for funding potentially disruptive technologies and may see them as a threat to be suppressed. However, it is essential to note that big business can become an ally once key disruptive technology shows promise. It is in early funding that big business is a tough sell.

Use of Capital

As we can see from the technology challenges facing our nation and world, we have a lot to do, and capital is limited. Hence, government grants and tax policies should prioritize the technologies listed above. Government policies that encourage private, primary investment into technologies developed by small businesses will be funding the "best and the brightest" and contributing to wealth growth and wealth distribution at the same time.

> Note: By "primary investments," we mean direct investment in companies. By "secondary markets," we mean creating securities that can be traded or sold to others. And, by "tertiary markets," we mean creating derivatives that can be traded or sold either based on the securities created or commodities bought or sold by the company.

A big problem in capital investing today is the high Return on Investment (ROI) expected by investors. All investors are "tainted" by the VC mentality recently developed in Silicon Valley – "triple my money in five years or less and get me out." This mentality tilts all investment strategies toward secondary and tertiary investment market considerations. The "game" creates a security that can be traded and leveraged. The actual ability of the technology to earn money is secondary to the "sizzle" and "sex appeal" of the "deal." It is all about appearances and "finance-ability." The question becomes, "How can we puff-up the deal, take it public and cash out?" It is much like planning a big wedding with the groom being only a minor consideration. Hence, scarce capital often attracts some of the biggest deals founded on risky technologies that "sound good enough" to be quickly "flipped" into secondary markets.

When capital financing becomes nothing but "bets" on financial "schemes," the chances for that capital creating sustainable sources of future wealth are diminished. The trading of securities created by the scheme can improve the wealth of some capital holders. Unless the businesses activities funded by raising capital are economically sustainable, they can negatively impact wealth distribution. Knowledge workers and innovators depend primarily on wages earned over time to build their wealth. When innovative workers invest their time in worthless schemes, they waste their precious and often irreplaceable human capital. Highly qualified individuals can be "conned" into investing much of their productive years in useless schemes only to be unemployed or underemployed in their later, more vulnerable years.

Another factor that comes into play is the government-supported diversion of capital to non-profits. Of course, tax breaks encourage individuals to put money into non-profits, but that is hardly the end of it. Strange collusions between government, the press, and universities have created almost a "shake-down" industry that coerces "big players" to "give" to favored non-profits "or else." Textbooks in business schools (even "conservative" ones) tout corporate "ethical" needs to be "doing good while doing well." Hence, there are many new pressures for disposable income to be "invested" in the financial "zero-sum-game" of non-profits.

From a financial perspective, capital "invested" in non-profits is a loser. There is no financial return except to the managers of the non-profits. Claims of economic benefits to society seem elusive and certainly difficult to quantify. Instead of creating and distributing wealth, capital disappears in "feel good" activities that frequently line the pockets of the ruling elite and often their relatives. Zero return on capital erodes the value of money and sinks it into economically unsustainable activities with insatiable appetites for more "investment." Dollars of known value "buy" transitory feelings and often empty promises of future bliss.

Taking disposable income out of the potential investment pool makes capital dearer, driving up the "cost of money," which translates into higher required capital returns for remaining available capital. This process makes the available capital more costly when compared to labor, eroding the "bargaining" position of laborers – even highly trained, professional "laborers." As Piketty points out, this further exacerbates the concentration of capital into the hands of the wealthy, facilitates exploitation, and foments social unrest. The growth of non-profits in the USA is remarkable. Piketty estimates that non-profit organizations held about 6% of the wealth in the USA in 2010 [6].

We should not miss the parallels between the wealthy, powerful, untaxed, religious institutions of the Middle Ages and the huge non-profits of today – especially the universities with billion-dollar endowments. Both benefit from an economically and politically weak labor force and the concentration of wealth and influence. Both make promises that are often undeliverable (at least in this life). Although economic data for the Middle Ages is scarce and unreliable, Piketty has reconstructed data from the eighteenth century that shows an alarming correlation. From 1750 to 1780, the Catholic Church owned approximately 25% to 30% of all property in Spain and about 25% of property in France [6]. These were times of slow economic growth, significant inequalities in the distribution of wealth, and the precursors to violent social upheavals. Thus, encouraging capital to grow non-profits seems a poor fiscal policy from an economic perspective.

To enhance the development of critical disruptive technologies, we need to encourage private, primary investments in emerging technologies. These should focus on small businesses – the engine of innovation. Furthermore, it will require changes in investment regulations, banking regulations, and tax codes. The target should be encouraging many small, private, primary investments of $250 K to $5 million.

Scarce Supply of Capital for Small Business

The supply of capital for small businesses has waxed and waned throughout US history. Since the banking changes that came out of the financial crisis of 2007–2009, money for small businesses has been very tight. The Dodd-Frank Wall Street Reform and Consumer Protection Act of 2010 effectively eliminated the "neighborhood bank" that directly supported local small businesses. The *Wall Street Journal* correctly predicted that this act would hurt start-ups, but the complete picture is much more complex.

A complex interaction of investment and banking regulations, tax policies, and increasing business risk in the USA drive small businesses' short supply of capital. For example, banking regulations have eliminated "reputation" lending – a common source of business start-up loans. In addition, banks now must collect data to support the ability to pay from current cash flows. These factors make it very difficult for a start-up to get a bank loan, especially if the technology development will take years. Unfortunately, the high priority Technology Development identified earlier will often require years of development.

Investment regulations increase the costs of financing transactions through the high cost of compliance. Rules promulgated by the Securities Exchange Commission (SEC) and individual states are a nightmarish mess that few, even investment attorneys, completely understand. Although the dizzying array of licenses and training requirements does little to reduce fraud and business failure, it does increase the cost of equity financing dramatically. Hence, only the biggest deals, with the quickest exit strategies and the best payouts, are likely candidates for equity financing.

Innovators need a reliable source of patient funding in the $250 K to $5 million range. An obvious funding source should be government grants, but this has not been effective. As we have seen, the selection processes for government contracting are very flawed. Furthermore, even the Small Business Innovation Research (SBIR) program supports operating businesses rather than start-ups developing disruptive technologies. A better, more effective approach would be to engage private funding. But, again, this will require significant changes in policies and regulations to encourage private funds to take on the long-term costs and risks of key disruptive technologies.

Despite contrary claims, many politicians and regulators have no interest in assisting small businesses. Frankly, they don't like small businesses. They are just "not their kind of people." Small business owners are too opinionated, independent, confident, self-reliant, and just plain too conservative – fiscally and morally. Small businesses are too hard to tax and control from a policy position. Many politicians view small business owners as pains in the butt not likely to fund their next election. To say the feelings are mutual would be an understatement. Until small businesses and politicians mend their relationships, government policies will continue to starve this engine of economic growth and technical innovation.

Accountability

Surprisingly few folks excel at managing Technology Development. I think the main drivers are:

- Few know how to evaluate technical performance and combine that with a realistic business plan,
- Many programs give a combined technical and business review a low priority,
- Many evaluators delude themselves for a wide variety of reasons, and
- There are few consequences for poor judgment or incompetence.

Though seldom used, the tools exist for practical evaluation. Unfortunately, many are unaware of how risk can be objectively measured and balanced against potential benefits. Our society is entirely ignorant of statistical inference, strangely uninterested in the topic, and skeptical of anyone who claims to know anything about it. Mark Twain quipped, "There are three kinds of lies: lies, damn lies, and statistics." Hence, many in positions of authority simply do not ask the question, "Where's the evidence for that?" They don't seem to expect much evidence, seem to know what they would do with the evidence, and don't seem to trust others to do anything useful with the evidence.

Hopefully, this is changing. Statistical software and powerful microcomputers are widely available. It is no longer necessary to subject students to the misery of statistical calculations for them to learn the basics of how to use statistics. Statistics is now used regularly in many scientific and medical fields. There is little excuse not to have a thorough, documented statistical analysis of technical data and projections made from them. Failure to require statistical analysis of claims made is bad government, bad business, and bad investment. The inability of many scientists, engineers, and business managers to understand basic statistical inference discredits our educational system. The general public seems flummoxed by anything related to chance, which is dangerous in a modern world where hardly anything is sure.

Changing our ability to deal effectively with uncertainty will take time and effort. The impetus to do so will not come from the charlatans who benefit from the flim-flam. It will not come from the fools who blame fate or some greater power. As a society, we need to incentivize statistical analysis. It will require intervention into our government contracting practices, our views of fraud and negligence, and our education system. We must come to expect a professional assessment of risk before we can effectively require it.

Patience

Effective development of disruptive technologies can take years – in fact, it is better if it does take years. Quick solutions are often no solution or fraught with terrible, unintended consequences. Having many potential solutions in the queue would be a wise strategy. One of the few positive outcomes of COVID-19 was realizing that several technical developments were in the queue and could be ready years before expected. We need to encourage more long-term research that could be the next "miracle." The critical question is, "How do we make risky, long gestation development practical?"

The COVID-19 experience may have lessons for us. The COVID-19 vaccines did NOT come from a government or academic lab. The politicians and PhDs were the ones that gave us the dire predictions that made it sound like "we're all gonna die." The solution came from private industry that put profits back into long-term possibilities. The "solution" did not require the CDC or the FDA to help out, but rather it required them to get out of the way. It was closer to being another example of a "Kelly Johnson Skunk Works" than a big win for the National Science Foundation or National Institute of Health.

Chapter 8
Where Do We Go from Here?

Taken altogether, we see that the private sector has the best ability to fund early, disruptive technology. Both governments programs and non-profits find it challenging to select technologies with a high probability of being economically sustainable. They simply do not have that as a mission or maintain the resources needed to succeed. Unfortunately, if we do not develop effective technologies, we could slip into an economic dystopia similar to the world described by Ayn Rand in Atlas Shrugged [26]. We could see significant challenges go unresolved, overall wealth shrink, and what wealth there is becomes more concentrated in fewer hands.

If we wish to improve our performance in developing vital disruptive technologies, we will need to encourage the private equity sector to invest in them. High priorities are:

- Reducing the influence of academics and higher education in Technology Development in government contracting. Apart from healthcare, the aim should be eliminating government set-asides to universities as soon as practical,
- Strengthening review of government Technology Development programs by reconstituting (not the same as reinstating) the Office of Technology Assessment (OTA),
- Specifically requiring statistically sound review of connections between the technology under development and business plans for all government-funded programs,
- Encouraging small, primary investments directly in small businesses engaged in disruptive technologies by exempting those investments from capital gains tax,
- Discouraging investments in secondary and tertiary markets by instituting a graduated capital gains tax,
- Immediately applying the graduated capital gains tax on non-profit endowment investment earnings – especially for universities and without regard for their association to other non-profit entities such as churches and hospitals,
- Phasing out tax-exempt status for all non-profits in 10 years and laying out rules for how a non-profit can become a for-profit entity,
- Phasing in strict financial reporting for all non-profits over five years with particular emphasis on public financial disclosures of executive salaries and complete executive compensation agreements,
- Reforming all of education to make it eventually competitive, for-profit with government funding by taxation distributed via vouchers to parents (K-12) and adult students (higher education),
- Requiring grade-appropriate probability and statistics curricula beginning no later than eighth grade,
- Requiring appropriate field proficiency in probability, statistics, and statistical inference for all STEM curricula,

https://doi.org/10.1515/9783110451634-008

- Requiring appropriate field proficiency in probability, statistics, and statistical inference for all professional certification in engineering, science, medicine, quality control, project management, investment, and finance, and
- Deregulating crowdfunding for small, Technology Development companies.

Reducing the Influence of Academics and Higher Education

As we have said earlier, the nexus between government and universities have had a detrimental effect on Technology Development – except perhaps in medical technology. They are frequently not experienced in business and have little interest in or ability to assess the practicality of technology. Furthermore, the academic community is a tangle of competing ideas, priorities, and fiefdoms with almost no awareness of the costs created by incessant deliberations. Dealing with universities is frequently excruciatingly slow and marked by irrelevant and debilitating turf wars and ego conflicts.

The aim would be to change higher education very dramatically. Our citizens have been impoverished with unforgivable debt while being given mediocre education. The academic environment is a medieval scholastic tyranny where facts, logic, and practicality don't matter. Academics ignore their customers (i.e., students and parents) and are fundamentally ill-equipped to think about serving a customer. Furthermore, the academic environment outside of the business schools is ignorant of and hostile to private industry. This environment needs to be changed to make it more effective in working with private companies and teaching students how to succeed in the private sector. Too many science and engineering students come to the end of their undergraduate education deep in debt and very anxious about how they will make money to pay off their debt. Many decide to go on to graduate school to postpone this day of reckoning. Unfortunately, many find that those graduate degrees leave them ignorant of and terrified by the private sector.

Example 33: My Experience with Universities (at least part of it)

I was one of those undergraduates that came to a crisis at the end of my bachelor's degree. I graduated from the University of Kansas with a bachelor of science degree in chemistry. I was an excellent student and won a couple of academic awards. I even had completed a couple of courses considered graduate-level and had been accepted by Iowa State University into a PhD. program in theoretical Chemistry. I had gotten married and was waiting to move from Kansas City to Ames when I decided I had had enough of academia for the time being.

I stayed on in my blue-collar job at a steel mill until I could figure out what to do. It was 1973, and the economy was in the deepest recession since WW2. It was the height of the oil crisis and the start of stagflation (high inflation and low growth). In 1975, I finally landed a chemistry job in an environmental lab. Out of necessity, I became an analytical chemist. It was challenging for a while. These were the early days of environmental compliance with many new analytical challenges. I quickly worked my way up into management and learned that I liked the business aspects of technology more than I had anticipated. In 1978, I decided to try my hand at running my own business.

I didn't have the capital to open a laboratory, but an opportunity in manufacturing came available. With my dad and my younger brother, we opened up a machine shop doing custom work and building a steel wire spooler that my dad had patented. My brother and I quickly learned accounting, taxation, finance, and management by reading and applying information supplied by the SBA. We could see that year upon year our business was getting more and more difficult. Our primary customer was the local steel mill, and the whole industry was struggling with rising costs and foreign competition. Finally, we decided to finish contracts, close the shop, sell equipment, and pay all debts. My brother went to dental school and became a very successful dentist in private practice. I went back into environmental chemistry.

In 1980, I moved to Colorado and took a job as general manager over a couple of fledgling environmental labs. This decision was the beginning of a series of job changes that moved my career toward the management of start-ups. I also began working on an MBA in finance and accounting. It had become clear that although I had an excellent education in Chemistry, my lack of business understanding was a severe handicap. Thus, I received my MBA in 1983 while working full-time.

Shortly after completing my MBA, I received a phone call from a chemistry professor at Ft. Lewis College in Durango, Colorado, about 350 miles away. The professor asked me if I knew anyone who could talk about alternatives to going to graduate school for undergraduate chemistry majors. His observation was that many of his students felt like they had to go to graduate school even though they were sick of school. He didn't think he could speak on the topic since it was what he had done as an undergraduate. He was hoping to find someone to give a talk in just a few days and was desperate. He had called all over Colorado, trying to find someone willing to speak on the topic. I was delighted to make the trip.

I learned at least two things from giving the talk at Ft. Lewis College. First, it allowed me to see that my experience was not unique. Many technical graduates feel unprepared for life by their university experience, and I committed myself to do what I can to solve that. Second, I learned not to be insulted by being the person of last resort – the one "at the bottom of the barrel." It is a great chance to exceed low expectations.

We should force universities to earn their money by teaching instead of camping out on government grants. They should compete for students and have limited ability to make money from sources other than education. We should judge professors and their universities on how well they prepare students for success in life. This change will not happen unless we bar them from filing patents to own technology that taxpayers, students, and parents have funded. We must change the way we fund education to transition universities from government-supported non-profits to for-profit businesses competing for students based on value to the student. The schools with the best deal will eventually win, and those with poor value will either change or die. These changes will take time, and the universities will be no help. It will require a revolution in higher education.

Of course, we must be cautious when privatizing university hospitals. We have been committed to delivering healthcare through many non-profit teaching hospitals. Given the bureaucratic nature of academia, I doubt that it has been an efficient delivery system, but it is what we have today. We cannot immediately dump it without catastrophic disruption to society. We will have to transition university hospitals to private hospitals, which could take many years. That should be done in a way to encourage fair competition. Hence, we must work out equitable disposal of university endowments. As a rule, we should distribute those endowments to create spin-off private hospitals, pay down student debt, fund patient care, and fund continuing research. We should ensure that these endowments do not become a windfall for executives and donors.

Strengthening Review of Government Technology Development Programs

We also need to improve our abilities to evaluate technology – especially in government contracting. This responsibility was once the role of the Office of Technology Assessment (OTA). Authorized by Congress in 1972, OTA became active in 1974. However, Congress defunded the OTA in 1995 because it had the reputation of being a bloated bureaucracy that did little good. The Government Accountability Office has partially resurrected OTA as the Science, Technology Assessment, And Analytics (STAA) Team. Congressional discussions around OTA and the STAA have long been a partisan issue. Democrats generally support this type of watchdog group. Republicans argue that OTA and STAA are still high-level boondoggles.

What is needed is a hardnosed combined technical and business analysis of the value and costs of technologies under consideration. All government contract selection processes should include significant weighting based on the economic viability for self-sustaining development. We should amend the Federal Acquisition Regulations (FAR) to require financial analysis of all federal technology development contracts using the potential return on investment processes described in this book. The new FAR clauses should force proposal reviewers to favor development plans with realistic development roadmaps that include evidence-based decision-making decision points. Some government contracting programs (e.g., SBA's SBIR program) include financial evaluation criteria, but the requirements are inconsistent and vague. As mentioned earlier, such programs suffer from using academics, government bureaucrats, and non-industry savvy consultants in the evaluation process. New FAR clauses should require selection reviewers to employ experts with specific industry knowledge. Unless resources – expertise and methodologies – are made available to government contractors throughout the entire contracting process, the effort will naturally be only a high-level Public Relations exercise with little real financial benefit.

A review of *GAO Science, Technology Assessment, and Analytics Team: Initial Plan and Considerations Moving Forward* (April 10, 2019) [28] and GAO's *Technology Assessment Design Handbook* (February 2021) [29] reveals some interest in the cost/value assessment, but only to the extent of reporting to Congress at a high level. What is missing is an "in the trenches" group that works across many agencies to assist many government Contracting Officers and review committees in finding high-value technologies to promote and eschew the flim-flams. A key question to ask would be, "Is this going to help prevent another Solyndra?"

Specifically Requiring Statistically Valid Review

A critical need in Technology Development is a statistically sound analysis of the technology and its connection with business planning. Although such analysis would be a powerful tool for balancing risk versus reward, it is a multi-disciplined task rarely done well. One of the barriers to adopting a statistically sound approach is the lack of generally accepted analysis methods. Various methods exist in many scientific, engineering, and medical fields, but the contracting process rarely requires their use. As a result, decision-makers frequently do not apply statistically valid financial reasoning for their decisions.

> Failing to require decision-makers to justify their decisions using statistically valid methods is an open invitation to a host of arbitrary, capricious, and self-serving acts that can result in inexcusable injustices and inefficiencies.

Another barrier to the adoption of statistically valid decision-making and review processes is the lack of consequences for making poor investment decisions – especially for government bureaucrats. Even in egregious situations, government employees rarely suffer any consequences for poor choices made in their jobs, including investing government money in absurd schemes. The practice of "absolute immunity" protects these individuals. The Westfall Act of 1988 expanded and strengthened that practice. Unfortunately, the approach tends to create moral hazard in that taxpayers frequently suffer the consequences of poor decisions made by bureaucrats with little hope of holding the individuals at fault responsible for their actions.

We could try to hold government employees responsible for big Technology Development blunders by pursuing criminal charges of fraud against them, but this is very hard to do. Prosecuting individuals for fraud is problematic because it requires proving intent to deceive for personal gain. It is always difficult to prove intent. It requires a level of "mind-reading" that is distasteful to legislators, regulators, jurors, and the general public. Let's face it, Americans can't help but sympathize with the clever con-artist if they don't cause too much damage, after all, "You can't cheat an honest man." [23] Furthermore, it is also unlikely that government employees will directly benefit from the scam. More often than not, we are trying to avoid a blunder and not a fraud.

One way to reduce moral hazard is to require decision-makers to share the risk of failure. We could do this most easily by shifting Technology Development from government to privately funded programs. There is rarely absolute immunity in the private sector, and individuals are prosecuted for fraud regularly. Therefore, we should privatize technology development where practical. We will talk more about that in a moment.

Nevertheless, many critical Technology Development programs of interest are high-cost, high-risk, long-term programs that are not likely to be funded without government assistance – especially in the early stages. Hence, if we could incentivize the adoption of better evaluation protocols, we would see better decisions made even in government-funded development. We mentioned earlier that adopting the processes proposed here into the FAR would improve evaluations. The critical question is, "Why would government agencies adopt FAR clauses that would tie them to an objective review of financial outcomes?" It seems unlikely that the federal government would adopt such a view. Nevertheless, I have seen at least two western states that focus public funds on job creation and launching sustainable businesses. Hence, we may see more success with state programs than federal programs – at least for a short term.

Encouraging Small, Primary Investments Directly in Small Businesses

Private investors do not yet have much incentive to make small, high-priority, high-risk, primary investments. Although the returns can be very high, the costs of finding, evaluating, and creating mechanisms for primary investments are high. Secondary and tertiary investments on existing, well-established exchanges are far less time-consuming and costly. Hence, small deals in primary markets are not very inviting. Unfortunately, investments in secondary markets do little to grow wealth and even less to improve wealth distribution. Following the logic of Piketty in Capital [5], creating new businesses with new technologies builds wealth. It creates economic mobility that enhances the value of labor (at least well-educated labor) and contributes to wealth distribution. One way to encourage these primary investments is to modify the tax code. An especially effective approach would be to allow big business tax breaks for investing in small business development of targeted technology.

Graduated Capital Gains Tax

Again, following the logic of Piketty, a graduated capital gains tax could encourage the kinds of private investments needed to develop critical technologies, grow wealth, and ensure that knowledge laborers have the opportunity to share in that wealth. We could modify Piketty's idea somewhat and make primary, small investments more lucrative by making the capital gains tax rate on returns on those investments low, perhaps even zero. A high Capital Gains rate on large investments could also re-direct investment money from secondary and tertiary markets that do little for laborers to selected primary markets that fund new operations in critical emerging technologies.

Capital Gains Tax on Non-profits

Non-profits should pay capital gains tax (including the graduated scale proposed above). For many decades the multi-billion-dollar university endowments have been exempt from capital gains tax on their investment returns. The Trump administration changed this in 2017 to a whopping 1.4%. This meager amount is much lower than the minimum 15% paid by for-profit companies. These unfair rates have multiplied the growth and influence of some of our most undemocratic institutions – universities and religious institutions. The missed revenue is not small potatoes from a tax revenue perspective since non-profits now make up the third-largest segment of our national economy.

Phasing Out Tax-Exempt Status for All Non-profits

To expand wealth and wealth distribution, we should tax many, if not all, non-profits. As pointed out earlier, growing non-profit wealth has been a bad fiscal policy. It has generally resulted in the concentration of wealth in the hands of a small, powerful elite. This negative fiscal effect is not some new revelation. Pope Innocent IV (Pope from 1243–1254 AD) set guidelines for the investment of Church wealth, setting aside the total prohibition against usury. Piketty notes:

> the problem was not usury as such; if usury yielded too much interest with too much certainty, however, the wealthy might be induced 'by avidity for profit, or to guarantee the security of their money,' to invest 'in usury rather than in less secure businesses.' The pontiff went on to cite as examples of 'less secure businesses' investments 'in livestock and agricultural implements,' good that 'the poor do not own' yet which are indispensable for increasing true wealth.
> Page 96 of [6].

Not only does the growth of non-profits take capital out of productive circulation and concentrate wealth, but it also dodges the essential questions about the money value of many services and obscures their actual cost. Thus, even if we loath putting a dollar value on human life, we dare not ignore the price for heroic treatment of terminal illnesses. Nor should we assume that we value saving homeless puppies over homeless people because dog lovers have a better donor base, more advertising, and more neighborhood organizers. We still must count the costs and balance them against benefits. Hiding costs by funding "free stuff" through lotteries, government subsidies, and non-profit distributions only obfuscates the cost/benefit relationship.

Hence, consumers should pay for everything, allowing the market to set money values. In the long-run, actual costs will set floors for various programs, and practicalities will limit funding. Furthermore, the governments should encourage free-market competition to set money values by consensus rather than fiat whenever practical. This approach has nothing to do with ethics. It is simply a matter of practicality. No regime is so infinitely knowing or powerful as to set or maintain a list of prices and values in any meaningful way.

Phasing in Strict Financial Reporting for All Non-profits

It is imperative to open non-profits to the scrutiny of standardized financial reporting as soon as practical. For generations, non-profits operated with almost no financial accounting standards. We expected them to eschew all filthy lucre. Their financial scruples were exemplary in many cases, but there have always been stunning examples of abuse. Today things seem much different. Today a running joke in business schools is, "If you want to make good money, go into finance. If you want to be rich, start a non-profit."

Example 34: Non-profit Marketing Projects

An associate professor of marketing at a conservative Christian university decided that students should gain real-world experience by doing marketing projects with companies outside the university. Students were placed into teams and encouraged to find marketing projects independently or with the professor's help. Several groups selected non-profits companies who were willing to participate with the students. Several of these had connections to the university and were very cooperative.

A consistent difference emerged between how for-profit and non-profit companies evaluated their products and services and designed their marketing strategies. The for-profit companies were naturally interested in customer desires and keenly interested in how their product or service stacked up against competitors. They hoped to give customers experiential and monetary value. Non-profits were generally quite unaware of competitors for their products or services and rarely considered how customers or users valued the offering. Their focus was on their donor's perceptions and psychological needs far more than the ultimate user's value.

Even more interesting, non-profits gave almost no consideration to the cost of delivering the products or services to the ultimate user. Their accounting information was seldom useful for cost analysis. Whatever the total cost was (including overheads) just became a part of the fundraising ask. There had also been virtually no thought given the existence of the higher value or complementary alternatives that might be available to ultimate users. In some cases, when students suggested working with other non-profits to reduce costs or add value for the user, their suggestions frequently met disdainful objections.

The non-profits also gave little thought to ultimate benefits to customers. Frequently, non-profits have little knowledge beyond a few anecdotal stories of who uses the products and services delivered. They were also woefully short on evidence about how effective the products and services met stated mission goals. In general, non-profit executives expected users and recipients to accept the product or service as intrinsically valuable with little evidence. Hence, customer feedback was rarely valued or wanted beyond the anecdotal video. Nevertheless, donor feedback on the product or service was crucial.

It became clear to the professor and many students that non-profits are focused on raising money and "selling a story" to donors. Increasing the donor base was the main purpose of marketing and was viewed as the primary source of wealth for the non-profit. The ultimate product or service value to the users was incidental.

Reforming All of Education to Make It Eventually Competitive, For-profit

Many of the problems we have in funding Technology Development are related to our educational system. Whether the topic is poor understanding of statistics, the need for technology, the scarcity of capital, the cost of money, the valuation of technology, or the impoverishment of young adults, education, is in the middle of the discussion. Furthermore, like so many ills in our culture, it seems that many of the most likely solutions will come through a much-improved educational system. Hence, it appears that an overhaul of our educational system is needed.

If we take what we have learned about non-profits, it seems reasonable to propose making education a for-profit endeavor. In other words, most of the problems that we see with non-profits appear in our academic institutions. It is not hard to see how education has become a bloated, out-of-date, self-serving behemoth in many ways. Unfortunately, that seems to be the tendency for all non-profits.

Privatizing education is not nearly as crazy as it may seem at first. There are already many for-profit educational institutions. Many of them are doing well financially (perhaps some more "well" than we would hope). Many schemes exist that allow tax dollars to flow to parents who would then have options for educating their students. Many of these are working very well. It seems logical to give parents and students more options and put pressure on schools to meet the needs of their customers.

Our educational system, which at one time was well ahead of the rest of the world, has fallen behind in a matter of only 50 years. It has fallen behind not because of underfunding. On the contrary, we spend more on education per student than many countries. We have fallen behind because education has lost sight of its mission. Primary and secondary education had the mission to make a literate nation. The educators knew that mission, and parents supported them in that mission. Even under some of the worst conditions, primary and secondary education succeeded in making a very high percentage of our populace literate.

After WW2, the game changed, but our institutions did not. We needed higher education to do more than teach teachers how to teach literacy. The universities needed to create more sophisticated curricula because we needed to build more complex machines, generate more energy, grow more food, and defend ourselves from enemies with more devastating weapons. Unfortunately, higher education never seemed to embrace such simple, practical goals. Instead, the universities peddled myths of happiness, fulfillment, enlightenment, and success to their donors and supporters while delivering none of these to students. As a result, in the 75 years since the close of WW2, we have become less happy, less fulfilled, less enlightened, and, for far too many of us, less wealthy. The same Pollyanna nonsense has now infected our primary and secondary schools. The mission creep of higher education was carried to the educational departments and eventually to our primary and secondary schools.

> One ironic success story for higher education is in sports. It has been challenging to promulgate the myth that happiness or enlightenment wins football games. The competitive nature of sports and the draw of big money to successful athletes have made many universities into excellent molders of professional athletes – much against the will of many "true academics."

Requiring Grade – Appropriate Probability and Statistics Curricula

We have listed here just some of the educational needs we have. There are many others, some of which are field specific, but none is more significant than our need to understand statistical inference. The fact that our government entities can dupe us into funding their schemes through lotteries shows just how bad we are at math. Sadly, they flimflam the most ignorant and needy among us.

We need to teach statistical inference beginning early. Students must experience statistical inference to understand how probability and statistics work. They must learn that failure is an option. We may improve our chances for success in several ways, but we can never eliminate the probability of a failure. Teachers should use games of chance, sports, science, and most activities that interact with a physical environment to show students how everything is uncertain and subject to probable outcomes. We need to think through how best to get our people to think in bets. That is, knowing odds, doing your best to improve the odds, and then taking a chance on a good bet without hesitation or future remorse. There is no reason to whine or seek justice when a good bet turns into a loss. It will happen. Just be wise enough and persistent enough to keep in the game. This kind of training needs to be started early and implemented in every corner of our society.

Requiring Appropriate Field Proficiency Probability and Statistics STEM and Professions

Many science, technology, engineering, and mathematics curricula include probability and statistics – every one of them should. But, unfortunately, probability and statistics training is universally bad. Instructors should save the theoretical mathematics for courses in advanced mathematics. Instead, the hands-on approach described above should be the primary training method. In addition, professional certification programs in the hard sciences, engineering, and business should require demonstrated working knowledge of probability and statistics.

Deregulating Crowdfunding for Small, Technology Development Companies

And finally, we would do well to deregulate equity crowdfunding. This funding method had much promise, but the SEC hamstrung it with regulations. Under the guise of consumer protection, the SEC has so loaded equity crowdfunding with rules, regulations, and liabilities that it has become almost useless. As a result, online platforms take too big of a chunk out of small investments to make equity funding of Technology Development viable. Recent changes raising the cap may make this nominally better, but it is still a waste of time.

The SEC has almost destroyed the value of equity crowdfunding by making it safe. There should be no attempt to make these small investments safe. The key should be in keeping investments small. Considerable effort should go into educating investors on how risky business is, assisting the small investor in finding good bets, and showing them how they can protect themselves from losing more than they can afford. It is obscene that we open casinos everywhere, run state lotteries, launch Internet gambling apps, and congratulate ourselves for now having sports gambling – all with payout odds designed to impoverish the average bettor. Yet, at the same time, the SEC crushes equity crowdfunding claiming that the government "is here to protect the consumer."

Chapter 9
Some Final Words

It has been my experience that the success or failure of a Technology Development project or program depends primarily on the character, knowledge, and skill of the executive that leads the effort. Nevertheless, the executive needs the help of many players to succeed. Therefore, we will speak to the executive first but then address investors, stakeholders, and citizens.

To the Executive

The Executive must learn to deal with the science of Technology Development. We cannot expect every entrepreneur to be a scientist or engineer. But the future belongs to those business leaders who have sufficient background to understand the basics of their technology and its nexus with making money. That begins with strategic and market issues and proceeds to the verifiable congruence between the technology and market opportunities. The successful executive **must** understand these scientific techniques well enough to manage their application. The executive has a fiduciary duty for which they should be held accountable.

Since Technology Development is dynamic, the executive constantly assembles, grows, and adjusts the organization to meet new challenges. Here the executive must consistently and artfully recruit, organize, motivate, evaluate and modify the Technology Development organization. Much depends on the ability of the executive to deal with difficult personalities and stressful conditions. The executive must persistently build a culture that drives results through all these challenges.

The executive must be committed to creating a disciplined organization that, through internalized values, naturally and effectively embraces both creative thinking and disciplined action. The organization should naturally abhor deception, dishonesty, and sloth, as well as bullying narcissists.

It may be possible for the executive to be a single renaissance person in small organizations. However, in organizations of any reasonable size, it is necessary to build an executive team with shared values and goals and similar knowledge bases. When the executive role is shared, the members of the executive team must be working together and not at cross purposes.

To the Technologist

The ivory-towered egghead is a dinosaur headed for extinction. The cost of education will not support the many layers, the massive salaries, and the wasted capital

https://doi.org/10.1515/9783110451634-009

much longer. A revolution is coming to education that will disenfranchise many technically trained educators. The educational system in the USA may just collapse under its own weight, but it doesn't much matter. We will need technical training, and we will get it from somewhere. If not the US universities, then from foreign ones.

New technologies, curricula, and logistics may replace most programs, professors, and educators. Schools and universities will no longer be allowed to force students to endure abuse and incompetence. The technologist of the future will be technically excellent AND will understand business. The new model will be like the NTU Renaissance program, where life/business skills come first and technical training later. Furthermore, training, especially technical training, will be a life-long activity.

The new technologist will learn new skills and even new fields throughout their career. If education reacts to this opportunity, they will find exciting new markets. The new educational system could support online expert training with constantly new curricula and different curricula. Tenured, on-campus positions will be rare and not highly valued. Half of the universities will close, and no one will miss them. Those that remain will have the opportunity to respond to the needs of their customers – the students – and establish life-long learning relationships with them. Furthermore, they could make good money by being a valued consulting/training partner with successful students in private industry and the government.

To the Investor/Stakeholder

The investors and stakeholders must be shrewd evaluators of technologies, markets, and leaders. Too often, the investors and stakeholders saddle excellent executives with impossible tasks. Investors and stakeholders should stop asking competent people to waste their lives on fool's errands. If there is no apparent connection between technology and a market, everyone should ignore the technology and find something useful. Investors and stakeholders should engage objective (if not cynical) evaluators of the technology under consideration before committing managed funds (private or public) in a Technology Development project. Failure to do so should be considered a serious breach of fiduciary duty to partners, supporters, and government agencies.

Investors and stakeholders also need to be wise judges of competence and character when selecting the executive(s) to manage a Technology Development project. Even promising technologies can fail to reach their potential if Executives are not up to the tasks at hand. Furthermore, decision-makers should focus on business merits first and investment returns second. Setting priorities focused on economic sustainability is a matter of ethics. When investors and stakeholders select executives and deals to make a quick buck, they kill wealth and contribute to inequality in wealth.

And finally, the investors and stakeholders need to ensure that the executive team is a functioning entity. Too often, investors and stakeholders interfere with the working of the executive team and create dysfunction. It is far too common for the executive team to accomplish very little due to internal conflict.

To the Citizen

And finally, let me speak to the general public. This short book has given an overview of good Technology Development. I hope that the public will insist that anyone developing a technology, making claims about a technology, or evaluating a technology will use these techniques. These techniques will minimize the probability of failure and fraud by holding claimants to account. If citizens can implement these processes, we could reverse the devaluation of money and the concentration of wealth.

We need to be especially harsh on those who misrepresent, obfuscate, or suppress evidence required to assess risks and benefits objectively. This harshness applies not only to financial investment schemes but also claims made by our regulatory bodies, politicians running for office, and elected officials. There is no excuse for allowing the damage done by false claims to go unavenged. We have financial and political fraud because we allow it.

At the same time, we should also demand that our educational system teach us the kinds of techniques needed to assess risks and benefits. We should insist that these be made widely available to technologists, business leaders, bureaucrats, and us. For the good of our society, we need to get schooled on how to "think in bets." Understanding statistical inference is more than just a financial issue. Our inabilities were on display during the COVID-19 crisis of 2020/21. We found it challenging to understand the information given to us by our scientists. We inferred conspiracy or fraud when confronted with the conflicting data that is certain to arise during any learning process. We need to understand these topics much better than we do now.

We should be especially harsh in our assessment of our educational system. We have invested heavily in primary, secondary, and higher education, yet there is nearly universal agreement that these institutions have failed us. We must insist that the student is the beneficiary of educational investments and that parents are engaged in the turnaround process needed to reform education. We, our children, and grandchildren need exceptional educational opportunities at affordable costs. There is a growing consensus that education will require public funding. Students and parents should be directly involved in redesigning our educational systems and ensuring that the investments meet student needs.

Armed with better information and tools to evaluate our investment opportunities, we should insist that the government get out of our way and let us invest directly into small deals that can help build our wealth. Moreover, we should be very active in reducing our dependence on "middle men" of all kinds that rob us of opportunities in the name of "protecting us" from risk.

We should be especially active in holding our government accountable for wasting our tax dollars on hair-brained schemes that line the pockets of their donors. We should be especially wary of "green scams," non-profits, lotteries, public/private partnerships, and any other program that entices or takes money from us without hope of long-term financial return. These devalue our money and often contribute to the already obscene inequality of wealth distribution.

The citizen should also look at the myriad of future technological challenges and ask, "Can centralized government planning make all, some, or even any of the right calls on these complex problems?" Unfortunately, centralized planners have a poor record on this score. I think a better question would be, "What policies and systems can a government implement to encourage innovation and commercialization of the new, disruptive technologies that we will need?" I doubt many will conclude that either Soviet-like central planning or *laissez-faire* capitalism is an optimal solution. It seems more likely that a mix of entrepreneurship and wise public policy encouragements makes more sense.

I hope that this short book has been a primer on managing Technology Development. Of course, it is impractical to say all there is to say in a single book. Nevertheless, I think I have included many of the most critical issues. I continue to be a student of this vital topic and value your feedback and experience on the matter. Feel free to contact me through my website, https://www.tek-dev.net. Good luck with this all-important cultural and economic mission.

Bibliography

[1] Ron Stites, The Art, Science and Discipline of Technology Development, 2015, Kindle eBook, Amazon.com,https://www.amazon.com/dp/B00WTMVP60/

[2] Peter Drucker, The Landmarks of Tomorrow, Harper, 1959.

[3] Jonah Berger, Contagious, Why Things Catch On, Simon & Schuster, New York, 2013.

[4] Clayton M. Christensen, The Innovator's Dilemma, Harvard Business School Press, 1997.

[5] Thomas Piketty, Capital in the Twenty-First Century, The Belknap Press of Harvard University Press, 2017.

[6] Thomas Piketty, Capital and Ideology, The Belknap Press of Harvard Press, 2020.

[7] Jacques Berlinerblau, Campus Confidential, Melville House, 2017.

[8] Annie Duke, Thinking in Bets, Making Better Decisions When You Don't Have All The Facts, Portfolio/Penguin, 2018.

[9] Wendell L. French, The Personnel Management Process, 5th ed., Houghton Mifflin Company, 1982.

[10] Robert Park, Voodoo Science, The Road from Foolishness to Fraud, Oxford University Press, 2000.

[11] Mark J. Anderson and Patrick J. Whitcomb, DOE Simplified, Practical Tools for Effective Experimentation, CRC Press, 2015.

[12] International Organization for Standardization. (2015), *Quality Management Systems – Requirements* (ISO standard no. 9001:2015).

[13] International Organization for Standardization. (2017), *General Requirements for the Competence of Testing and Calibration Laboratories* (IS standard no. 17025:2017).

[14] David Amis and Howard Stevenson, Winning Angels: The Seven Fundamentals of Early-Stage Investing, Pearson Education, London, 2001.

[15] Damiano Montani, Daniele Gervasio and Andrea Pulcini, "Startup company valuation: The State of art and future trends," International Business Research, Vol. 13, No. 9, August 12, 2020.

[16] Pierre M. Gy, Sampling of Heterogeneous and Dynamic Material Systems, Theories of Heterogeneity, Sampling and Homogenizing, Elsevier, 1992.

[17] Steven K. Thompson and George A. F. Seber, Adaptive Sampling, John Wiley & Sons, 1996.

[18] Steven L. Brunton and J. Nathan Kutz, Data-Driven Science and Engineering, Machine Learning, Dynamical Systems, and Control, Cambridge University Press, 2019.

[19] Howard Gardner, Creating Minds, An Anatomy of Creativity Seen Through the Lives of Freud, Einstein, Picasso, Stravinsky, Eliot, Graham, and Gandhi, Basic Books, 1993, 2011.

[20] Martin Coe, Robust Systems Engineering for Medical Device Systems, Martin Coe, 2019, https://www.amazon.com/Robust-Systems-Engineering-Medical-Device/dp/1703701097.

[21] Arthur Pearl and Frank Riessman, New Careers for the Poor, The Free Press, Collier-Macmillan, 1965.

[22] Claire Dunning, "How the rise of urban nonprofits has exacerbated poverty," The Washington Post, September 24, 2019. https://www.washingtonpost.com/outlook/2019/09/24/how-rise-urban-nonprofits-has-exacerbated-poverty/

[23] Quote from Mordecai Jones, "The Flim-Flam Man," 20th Century Fox, 1967 film.

[24] John Mandel, The Statistical Analysis of Experimental Data, Dover Publications, Inc., 1964.

[25] Kim H. Esbensen and Brad Swarbrick, Multivariate Data Analysis, 6th Edition, An Introduction to Multivariate Analysis, Process Analytical Technology and Quality by Design, CAMO Software AS, Oslo, Norway, 2018.

[26] Ayn Rand, Atlas Shrugged, Penguin Group, New York, 1957.

https://doi.org/10.1515/9783110451634-010

[27] Nassim Nicholas Taleb, Fooled by Randomness, The Hidden Role of Chance in Life and in the Markets, 2nd ed., Random House, New York, 2004.
[28] Government Accounting Office, Science, Technology Assessment, and Analytics Team: Initial Plan and Considerations Moving Forward, April 10, 2019. See: https://www.gao.gov/assets/2020-02/GAOScienceTechPlan-2019-04-10_0.pdf
[29] Government Accounting Office, Technology Assessment Design Handbook, February 2021. See: https://www.gao.gov/products/gao-21-347g

Index

https://doi.org/10.1515/9783110451634-011

www.ingramcontent.com/pod-product-compliance
Lightning Source LLC
Chambersburg PA
CBHW061213240926
51846CB00006B/1209
9783110451719